Dante's Visions

I0888346

Dante's Visions: Crossing Sights on Natural Philosophy, Theory of Vision, and Medicine in the Divine Comedy and Beyond offers a fascinating insight into Dante's engagement with the science of his time, particularly with visual perception and neurological disorders. The relationship between the soul and the body and the bond between human beings and their natural environment were significant areas of interest in the medieval world. In Dante's *Divine Comedy*, as well as in his *Vita Nuova* and *Convivio*, these connections are enhanced to the fullest, expressing feelings and sensations, pain and ecstasy, and physical and spiritual passions under exceptional psychological and environmental stimuli.

Based on the research of a multidisciplinary group of scholars – including experts in Dante, the culture and history of medieval literature and philosophy, historians of science, neuroscientists, and specialists in vision and visual illusions – this book explores the poet's psychophysical descriptions of sense perception, the theory of vision, optical illusions and deceptions of sight, neurological phenomena, and the anatomy and physiology of the human nervous system. It highlights the Aristotelian sources of his scientific culture and the influence of the Arabic sciences on their dissemination in the Western world.

In addition to illustrating the cultural background of a poetic genius, with specific reference to the rich scientific reflections in Italy at Dante's time, this book brings out the many opportunities for future research at the intersection of science and literature in the past.

Cecilia Panti is Professor of History of medieval philosophy at the University of Rome Tor Vergata (Italy). Her research interests in the fields of philosophy, optics, theory of music, and the quadrivium are featured in numerous academic journals and collective volumes. Her publications include editions of Robert Grosseteste's cosmological treatises (2001) and his *De luce* (2011); Johannes Tinctoris' *Dictionary of Music* (2004); a volume on acoustics in Late Antiquity and the Middle Ages (2008); and the special issues of the journals *Micrologus: Nature, Sciences and Medieval Societies* ('Latin and Arabic Theory of Perspective', Volume 29, 2021) and *Revista*

Española de Filosofía Medieval ('Robert Grosseteste and Aristotelianism', Volume 30/1, 2023).

Marco Piccolino has taught general physiology and the history of science at the University of Ferrara (Italy), where he is currently a member of the Centre of Neuroscience. He is a neurophysiologist who has conducted significant research in the physiology of the retina, publishing his results in prestigious international journals, including *Nature* and *Science*. He has written several volumes on the history of electrophysiology and sensory physiology, notably *The Shocking History of Electric Fishes: From Ancient Epochs to the Birth of Modern Neurophysiology* (2011) with Stanley Finger; *Shocking Frogs: Galvani, Volta, and the Electric Origins of Neuroscience* (2013) with Marco Bresadola; and *Galileo's Visions: Piercing the Spheres of the Heavens by Eye and Mind* (2013) with Nicholas J. Wade.

GPHNP

GLOBAL PERSPECTIVES ON THE HISTORY OF NATURAL PHILOSOPHY

Series Editors:
Yael Kedar (Tel Hai College, Israel), Cecilia Panti (University of Rome Tor Vergata, Italy) and Nicola Polloni (University of Messina, Italy)

Fostering a new approach to the study of the history of natural philosophy this series aims to expand the discussion on natural philosophy cross-culturally and comparatively by focusing on the philosophical reasoning about nature developed particularly, but not exclusively, in three main cultural settings: Europe, the Middle East, and China. One of the main focal points of *Global Perspectives on the History of Natural Philosophy* is the interplay between philosophical and scientific concepts, stances, and problems arising from the premodern consideration of nature, broadly considered. Accordingly, the series provides a cutting-edge framework in which natural philosophy can be considered from new philosophically meaningful angles. Acknowledging the historical interweaving of philosophy and science of nature, the series publishes monographs and edited volumes dealing with the history of natural philosophy from three methodological perspectives: philosophical analysis, historical reconstruction, and comparative studies. Submitted manuscripts may either examine authors and issues from a specific philosophical tradition or engage comparatively with patterns and problems shared by different cultural settings.

Theories of Colour from Democritus to Descartes
Edited by Véronique Decaix and Katerina Ierodiakonou

Dante's Visions
Crossing Sights on Natural Philosophy, Theory of Vision, and Medicine in the Divine Comedy and Beyond
Edited by Cecilia Panti and Marco Piccolino

Dante's Visions

Crossing Sights on Natural Philosophy, Theory of Vision, and Medicine in the Divine Comedy and Beyond

Edited by
Cecilia Panti and Marco Piccolino

Routledge
Taylor & Francis Group
LONDON AND NEW YORK

First published 2025
by Routledge
4 Park Square, Milton Park, Abingdon, Oxon OX14 4RN

and by Routledge
605 Third Avenue, New York, NY 10158

Routledge is an imprint of the Taylor & Francis Group, an informa business

British Library Cataloguing-in-Publication Data
A catalogue record for this book is available from the British Library

ISBN: 978-1-032-53556-2 (hbk)
ISBN: 978-1-032-53557-9 (pbk)
ISBN: 978-1-003-41260-1 (ebk)

DOI: 10.4324/9781003412601

Typeset in Sabon
by codeMantra

Contents

List of contributors *ix*

Introduction 1

1 Dante and Natural Philosophy 8
SIMON A. GILSON

2 Visual Motion Illusions in the Classical Era and in the Middle Ages 29
NICHOLAS J. WADE

3 Vision as a Tangible and Dynamic Tool in the *Divine Comedy*. An Overview 45
MARCO PICCOLINO

4 Moving Clouds and Bending Towers: The Illusive Motion of the Garisenda in *Inferno* XXXI 62
MARCO PICCOLINO

5 Where Do Visions That Do Not Come from Sight Come from? 84
MIRKO TAVONI

6 Seeing the Light: Dante and the Perspectivist Theory of Light as Proper Visible 105
CECILIA PANTI

7 Visual Perception in Dante's *Commedia* According to the Early Commentaries (1320–1400) 126
FRANCESCA GALLI

8 **A Medical Commentary on the *signa amoris* in the *Vita Nuova*** 149
FRANCESCO BRIGO

9 **Dante, Healthcare and Diseases** 166
MICHELE A. RIVA AND LORENZO LORUSSO

10 **Dante, Neurologist and Neuroanatomist: Evidence from the *Divine Comedy*** 190
DONATELLA LIPPI, RAFFAELLA BIANUCCI, ELENA VAROTTO, FRANCESCO ARBA AND FRANCESCO MARIA GALASSI

Index 203

Contributors

Francesco Arba, MD, PhD, is a clinical neurologist and stroke physician at Careggi University Hospital, Florence. Besides his clinical duties, he is involved in clinical research projects and collaborates as a forensic neurologist for Italian courts. He has a major interest in the history of Neurology.

Raffaella Bianucci, MSc PhD, Université Paris-Saclay, is a bio-anthropologist and paleopathologist. She has an extensive research portfolio involving physical anthropology and paleopathology, mummy studies, pathography, and iconodiagnosis. She co-edited the book *Ars Medica nelle Gallerie degli Uffizi* (2022) and she is member of the Board of Directors of the International Society of Iconodiagnosis.

Francesco Brigo, MD, is a clinical neurologist and epileptologist with a deep interest in the history of medicine and neuroscience. He has authored numerous articles on historical topics and currently serves as Chair of the Study Group on the History of Neurology within the Italian Society of Neurology.

Francesco Maria Galassi, MD PhD, is Associate Professor of biological anthropology at the University of Lodz (Poland). He is also Adjunct Associate Professor in the School of Biomedicine, The University of Adelaide (Australia). His research areas cover the history of medicine, palaeopathology, and biological anthropology.

Francesca Galli is a postdoctoral fellow at the University of Zürich, funded by the Swiss National Science Foundation (Ambizione Grant 201935), and teaches medieval literature and philology at Università della Svizzera italiana. Her interdisciplinary research explores the intersections between medieval sciences, particularly optics, and literature, as well as mendicant preaching and the relationship between texts and images. Her first monograph, based on her PhD dissertation, is devoted to Bartolomeo da Bologna's *De luce* (2021).

Simon A. Gilson, FBA, is Agnelli-Serena Professor of Italian at the University of Oxford and fellow of Magdalen College. His publications include

Medieval Optics and Theories of Light in the Works of Dante (2000), *Dante and Renaissance Florence* (2005), and *Reading Dante in Renaissance Italy: Florence, Venice and the 'Divine Poet'* (2018).

Donatella Lippi is Full Professor of History of medicine at the University of Florence. Responsible for the Medici Project, she has authored more than 400 scientific publications and superintended several exhibitions in Florence and Germany. Her innovative didactic methodology grounds on Visual Thinking Strategies to improve the clinical and artistic eyes of the future doctors. She is President of the Lyceum Club Internazionale of Florence and a member of the Board of Directors of the Fondazione Santa Maria Nuova and Fondazione Careggi of Florence.

Lorenzo Lorusso is Director of the Neurology and Stroke Unit at Merate Hospital, Merate (Lecco). He is the author of numerous articles in international journals in neuroimmunology, neuro-oncology, and history of medicine.

Cecilia Panti is Professor of History of medieval philosophy at the University of Rome Tor Vergata. Her research interests concern medieval and early modern natural philosophy, optics, theory of music, and the quadrivium, with numerous publications in international journals and collective volumes. Her books include the critical edition of Robert Grosseteste's *De luce* (2011) and the editorial care of the special issue *Arabic and Latin Theory of Perspective* for the journal *Micrologus* (2021).

Marco Piccolino taught general physiology and science history at the University of Ferrara. He is member of the Centre of Neuroscience at the same University. As a neurophysiologist, he has carried out important research in the physiology of the retina by publishing his results in the main international journals, among which *Nature* and *Science*. He has written several volumes on the history of electrophysiology and sensory physiology, which are reference books in the field.

Michele A. Riva, MD, PhD, is Full Professor of History of medicine at the University of Milano-Bicocca and Director of the Department of Occupational Health of the Fondazione IRCCS San Gerardo dei Tintori in Monza. He is currently serving as National Secretary (Italy) of the International Commission on Occupational Health. His main fields of research are History of Occupational Health, History of Neuroscience, Paleodemography and Paleoepidemiology, and Medical Humanities.

Mirko Tavoni is Professor Emeritus at the Università di Pisa, where he has taught History of the Italian Language and Dante studies. Ordinary member of the Accademia della Crusca and other prestigious institutions, he has written extensively on Dante, including the edition of the *De vulgari eloquentia* (2011) and the volumes *Qualche idea su Dante* (2015) and edited with Bernhard Huss, *Dante e la dimensione visionaria tra Medioevo e prima età moderna* (2019).

Elena Varotto, PhD, is lecturer at Flinders University, College of Humanities, Arts and Social Sciences. She also works as a forensic anthropologist for Italian courts. She is Co-Director and Co-Founder of the FAPAB Research Center (Sicily, Italy).

Nicholas J. Wade is Professor Emeritus of Psychology at the University of Dundee, Scotland. His research interests are in the history of vision research, binocular and motion perception, and the interplay between visual science and art. He has written books on these topics, the most recent of which is *Vision and Art with Two Eyes* (2023).

Introduction

> The future of Dante studies will be dominated by the natural sciences, when these will become sufficiently refined and develop a capacity for thinking in images.
>
> **Osip Ėmil'evič Mandel'štam,** *On Dante*[1]

This volume stems from the seminar "Dante and the Neurosciences" that Marco Piccolino and Nicholas J. Wade organised as part of the 26th Annual Meeting of the International Society for the History of the Neurosciences held in Rome in 2022 (14–18 June) as a tribute to the Dante Alighieri celebrations running in that year. The seminar focused on the Florentine poet's depictions of physiological and psychological phenomena and pathologies, aligning them with contemporary scientific investigations into the nervous system, and on his sight theory, comprising the issue of optical deceptions. As a result of the great interest aroused by the seminar and the subsequent discussions among the participants, who included physicians, historians of science, and humanists, the need arose to collect the contributions within a book that would explore these themes in greater depth, taking advantage on the interdisciplinary insights emerged on that occasion.

The present volume is the result of this challenging initiative. It includes refined versions of some of the papers presented at the seminar and new studies commissioned from Dante specialists who kindly participated in this undertaking. We are profoundly grateful to all the authors for their valuable collaboration and dedication, which allowed this volume to include additional areas of investigation. We have ideally gathered the essays under the title of *Dante's Visions*, which, in our intention, signify the multifarious aspects of Dante's poetical narratives concerning the vision of the world, both perceptive and scientific, the theory and physiology of eye vision, the self-reflecting vision of internal moods, pathologies and psychological events, and the vision of the human body. These interests emerge everywhere in his writings, and this volume points the attention on some exemplifications of them, which may be grouped into three main disciplinary areas, namely, natural philosophy, theory of vision, and medicine; nonetheless, this book aims to offer an interdisciplinary

DOI: 10.4324/9781003412601-1

approach on these scientific perspectives, including literary and science history, history of philosophy, vision physiology, and neurology.

Our decision to focus on Dante and these topics has a specific motivation, which transcends the original celebrative occasion. Dante is among the most eminent medieval representatives of an idea of knowledge open to all human beings, which he seeks to disseminate through poetry in the vernacular. He sticks to this task by assuming the human being not only as the knowing subject but also as the main object of knowledge. His poetical investigation starts with the 'self', and specifically with the role of sense perception, and vision in particular, as an instrument of knowledge of the external world and the inner world. Thus, Dante's visions sweep in a wide variety of observations, real and imaginary, and representations of feelings, translated within the poetical language, which becomes a web of physical and psychological manifestations of consciousness that the poet communicates to the reader of his verses, both contemporary and future, offering an incredibly rich stimulus for our understanding of the medieval mind and world, and of our nature as living beings.

Dante explores the self-exegesis of the lyrics celebrating his passionate love for Beatrice in his first literary work, the *Vita nova* (c1295). Here, Beatrice arouses the psychophysiological reactions of the poet, emphasising the uniqueness and novelty of amorous passion, which subvert the normal physiology of perception, feeling, and cognition. The exploration of the feeling of love paves the way to a wider sense of 'love'in Dante's subsequent, unfinished prosimetric enterprise, the *Convivio*. It opens with a highly significant statement by Aristotle: "As the Philosopher says at the beginning of the *Prime Philosophy*, all human beings by nature desire to know". The poet responds to this authoritative assertion by waitering his banquet with the offer of the "meat" of his verses "on love and virtue" accompanied by the "bread" of their exegesis, that "will be the light that renders visible every shade of their meaning". Here, the feeling of love as search for virtue is inspired by the Donna Gentile, personification of philosophical knowledge, acting as a consoler after Beatrice's passing away. The new love traces the path of a wide range of explorations in different disciplines, in search of a knowledge directing the human reason to goodness. Concordantly, Dante's philosophical poetry undergoes an impressive refinement of poetical language, style, and ideas, offering us solid evidence of a keen interest in the mechanism of the world, in the theory and physiology of vision, and in the exploration of the human mind and psyche.

Dante's poetical masterpiece, the *Commedia*, opens our research on Dante's visions on a much wider scenario. Here, Dante's chain of tercets works as a self-sufficient language and personal narrative style to popularise the pilgrim's experiences of knowledge in the bitter years of his wandering "per l'altrui scale" and in his visionary journey from Hell to Heaven. In narrating his ascension from the blindness and darkness of ignorance and sorrow to the full vision and light of true knowledge, Dante's poetical awareness is so acute that he can 'touch' philosophy, politics, science, theology, mysticism, and all aspects of human understanding without being compelled to explain which sources inspire his evocative world, why he chooses one or another term, historical figure, metaphor, syntagm, or rime, or which stimulus, reading, or experience it conceals. In this evocative context, Dante's scientific knowledge underlies his expression of feelings and perceptions, pain and ecstasy, and physical and spiritual passions suffered under exceptional psychological and environmental stimuli. The *Inferno*, in particular, presents us with cases of nervous diseases, such as epilepsy or narcolepsy, as well as visual disorders, loss of consciousness, hallucinations, and several other symptoms that affect the damned souls and, sympathetically, the poet. The *Purgatorio*, in turn, presents cases of sensory deceptions and episodes of fainting due to unexpected emotions, while, in the *Paradiso*, Dante makes an extraordinary effort to explain the mechanism of perception, especially vision and hearing, under exceptional physical and psychological conditions.

The opening essay of the volume is by *Simon A. Gilson*, who points our attention on Dante's interest in Aristotle's writings on natural philosophy and the ways the poet redirected them in his writings, above all the *Convivio* and the *Commedia*. Gilson underlines how at Dante's time this part of the Aristotelian corpus of writings had been enriched, and complicated, by a substantial body of scholastic "literature", that Dante assimilates and transforms. After discussing the more relevant recent approaches in this area of Dante studies, Gilson concentrates on examples showing the pervasive assimilation of Aristotle and his commentators, especially Averroes and Albert the Great, drawn especially from meteorology. This science explores the sublunar phenomena due to the four elements: earth, water, air, and fire. Ranging from the depths of the earth to the upper regions of the sky, medieval meteorology was basically a science of vision, based on the observation of earthly exhalations, volcanos, tides, and the "upper impressions" of the spheres of air and fire, such as comets, haloes, the rainbow, and the Milky Way. Gilson authoritatively explains how Dante uses these phenomena "for a variety of expressive and narrative purposes, often in complex, extended sequences of imagery that distinctively shape the diegesis and leave an imprint on the poem's language and style". The last part of the essay explores the impact and relevance of the Dominican theologian and polymath, Albert the Great, whose magisterial systematisation of Aristotle Dante appreciates and widely employs, even more than Aquinas' Aristotelian commentaries.

With the second essay, by *Nicholas J. Wade*, we enter in an area of visual experiences that is the topic of other contributions collected here, that is, motion illusions. These kinds of optical deception have been recorded since the classical era, and Wade reconstructs the main scientific insights concerning these phenomena up to the Middle Ages. As an authoritative specialist in science history and optics, he refers to a wide variety of examples taken from past Greek, Arabic and Medieval science with a special attention to the role of moving clouds and water flowing, as mobile natural elements that display an appreciation of the relativity of motion perception with respect to the frames of reference that are in operation.

The two following essays are by *Marco Piccolino*, who emphasises two tightly related aspects of Dante's great interest in optics and vision. The first essay points to the pervasive presence, in the *Commedia*, of episodes in which either visions or terms connected to the act of seeing are used for suggesting tactile experiences, narratives of movements, and what has been called the 'cinematographic' perspective of the poem. This overview serves to introduce the second essay by Piccolino, which deals with a particular type of visual illusion narrated in canto XXXI of the *Inferno*, the apparent fall of the Garisenda, one of the leaning towers of Bologna. Dante adopts the sense of fear in perceiving the apparent tilting of the tower to the viewer when a cloud passes over it as a simile to express Dante's trepidation in looking at the giant Antaeus next to bending over the two poets. Dante seems the first to have described this kind of illusive motion referred to a terrestrial building, as a perceptual consequence of the real movement of a cloud. The second part of the essay delves into the scientific and historical aspects of Dante's Garisenda observation by presenting how past and recent commentators of the *Commedia* have referred to the episode.

The essay by *Mirko Tavoni* enters in another fundamental topic concerning Dante's visions, namely how the poet conceives of the origin of images and visions that are produced in the mind without coming from the sense of sight. In the first part of the study, Tavoni examines verses and passages from the *Vita Nova* concerning the phenomena of 'inner vision', which grounds the experience of love for Beatrice. Then, he proposes an interpretation of *Convivio* II viii 13 concerning dream divinations ("le divinazioni de' nostri sogni") and *Purgatorio* XVII 13–16 on imagination ("O imaginativa che ne rube talvolta sì di fuor"), which is discussed in the context of Aristotelian science, with special reference to the works of Albert the Great and Avicenna. Tavoni emphasises that the role of imagination undergoes a significant evolution from the early *Vita nova* to the subsequent *Convivio* and *Commedia*, thanks to the authoritative sources of Albert and Avicenna. Specifically, for Dante these psychic phenomena are not pathological signs of a weak personality but, on the contrary, "are privileged phenomena, reserved for strong personalities, of great nobility of soul, and endowed with prophetic value". The prophetical background of the inner vision is very likely a strong encouraging tool for Dante "to launch himself into writing the poem that would give shape to the total vision of the three kingdoms of the afterlife".

The following contribution is by *Cecilia Panti*, who focuses on another aspect of Dante's account of vision, namely the role of light as the main object of sight. She proposes a summary historical reconstruction of the tenet of light as proper visible, evidencing that this concept stands apart from the two prevailing optical theories followed at Dante's time – namely the Neoplatonic extromissionist thesis of the emission of visual rays and the Aristotelian intromissionist thesis of the reception of forms in the eye. In both theories, colour alone is the proper visible. Instead, the idea that light is something sensed per se by sight stems within the context of the recent Western reception of Alhazen's *Perspectiva*. Dante sticks to this principle since his early writings, and this, consequently, challenges his alleged early adhesion to the extramissionist theory. Panti underlines that Dante was neither a scientist nor had scholarly interests in divulging new and original optical theories; instead, his adhesion to this innovative tenet of contemporary perspectivism was strategic to the employment of light as a potent source of gnoseological, theological, and poetic metaphors, some of which are cursorily presented as exemplification in the last section of the essay.

The essay by *Francesca Galli* concludes the different approaches to the topic of Dante's visions in this volume. She presents how the first readers and commentators of the *Commedia* – dating from the period between Dante's death and the beginning of 1400 – 'visualized' and interpreted Dante's passages addressing optics and visual perceptions. Galli shows that at least six of the first commentaries, namely those by Iacomo della Lana, Benvenuto da Imola, Francesco da Buti, the *Ottimo Commento*, the *Amico dell'Ottimo*, and the Anonimo Fiorentino, refer to *perspectiva* as a science useful for illustrating verses and episodes set in hell (particularly from *Inf.* IX, X, XV, and XXXI), the realm where the eye is not dazzled, but rather tries to grasp something in the darkness. In addition, Galli examines how numerous hints about the theories of vision in these commentaries are concerned with the extramissionist and/or intromissionist stances, the anatomy of the eye, and the various factors of visual perception that can mislead the sense of sight. The wide scenario of these early notes of reading, mostly explored here for the first time, reveals how rich is the early reception of Dante's masterpiece also with regard to the topic of vision.

The last three contributions of this volume explore Dante's vision concerning medicine. They all are presented by professional physicians.

Francesco Brigo's investigation of the *Vita Nuova* concerns Dante's use of the medical theory of the spirits to describe the effects of the vision of Beatrice on the poet's body and soul. Brigo underlines that Dante enriches his examination of human physiology by emphasising the role of the brain in explaining cognition for conscious perception of external stimuli. Yet, Brigo insists on the relevant interplay between vision, dreams, memory, ideation, and imagination, emerging from Dante's detailed descriptions of the symptoms that precede fainting due to an intense emotional reaction. Dante's accounts are rigorous and consistent with contemporary medicine; however, the physiological and psychosomatic reactions to Beatrice's sight, and also without directly seeing her, do not follow the *ordo naturalis* expected by the medical theories of the time. Intriguingly, Dante places an overemphasis on the bodily effects due to the vital spirits of the heart, which arouses exceptional psychophysiological reactions due to the uniqueness and novelty of the love experienced by the poet. The supernatural event of Beatrice's apparition subverts the normal physiology of perception, feeling, and cognition with the miraculous power of attracting the spirit of the poet towards the place where his desires will be eventually fulfilled.

The essay by *Michele A. Riva* and *Lorenzo Lorusso* focuses on the role of medicine as a component of Dante's knowledge. The authors investigate three areas of medical inquiry. First, the advances and discoveries in the anatomical field. Dante displays keen observational skills in describing several symptoms of illness and diseases, and he is often able to relate health problems to specific organs and systems. For instance, the syncope as an expression of cardiovascular diseases; narcolepsy and epilepsy associated with alterations of the nervous system; voice disorders caused by laryngeal and pulmonary disorders. The Italian poet also reported infectious diseases such as scabies, malaria, and leprosy. His accurate description of these pathologies as effects of the punishments inflicted in Hell demonstrates observation skills that are still essential today in clinical diagnosis. The authors suggest that Dante might have conducted direct observations of numerous pathologies as a consequence of his active involvement in public and military events, which may have put him in contact with institutions where sick or wounded people were hospitalised.

Donatella Lippi, *Raffaella Bianucci*, *Elena Varotto*, *Francesco Arba*, and *Francesco Maria Galassi* close the present volume with an essay on Dante's references to neurological diseases. For the authors, the poet shows familiarity with fundamental aspects of functional neuroanatomy, that can be grasped in his descriptions of the symptoms and signs associated with various neurological diseases. Among them, the essay considers the spasmodic torticollis, named "Manto Syndrome" exactly after Dante's association of it with the damned fortune-teller Manto, and other pathologies of the nape region, which proves to be central in the poet's adhesion to the Galenic theories on the relevance of the brain. Dante reaffirms the neurological importance of this anatomical zone in several episodes of the Hell. The authors eventually refer to the recently advanced hypothesis that Dante himself might have suffered from neurological syndromes, specifically the fight-or-flight response.

In concluding the introduction to this volume, we wish to emphasise the great pleasure and professional advantage of having collaborated with renowned scholars and scientists in our explorations of Dante's visions. As editors of this collective book, we are fully conscious that several other perspectives of analysis of Dante's visions remain out of our proposal. After all, the never-ending historiography on every single brick of knowledge and emotion that our poet lay in his magnificent buildings excludes since the very beginning any possibility of completeness. Yet, each contributor to this volume, both the specialists of Dante and the scientists and humanists who are less familiar with him, perceives that Dante's world is a personal synthesis. Surely, it is not only a synthesis of the knowledge of his time that learned laymen, rather than an intellectual elite, think to be worth addressing, but also a synthesis that his poetry offers as a datum, in pretending a face-to-face dialogue between the poet and the reader. This is perhaps why Dante's visions are so profoundly rooted in his time and yet persistently present and living along the centuries, and today.

Note

1 Mandel'štam, Osip Ėmil'evič. *Sulla poesia*. Milan: Bompiani, 2003, 133 (our translation).

1 Dante and Natural Philosophy

Simon A. Gilson

1 Dante and Medieval Aristotelianism

A characteristic feature of Dante's writings is his manifold interest in a philosophical vision of nature and its processes, one based in large part on the recovery of what were known in medieval university curricula as the *libri naturales*, that is, Aristotle's writing on natural bodies in motion, the *Physica* (*Physics*), on the four elements and how they combine, the *De generatione et corruptione* (*On Generation and corruption*), on the heavens, *De caelo et mundo* (*On the Heavens*), on weather and its effects, the *Meteorologica* (*Meteorology*), on the human body, the senses, and soul (*On the Soul* and the short psychological treatises that make up the so-called *Parva naturalia*), as well as works on plants, minerals, and the animal kingdom.[1] By the late 13th century, the corpus of Aristotelian natural philosophy, with related conceptual and methodological tools, had been supplemented, enriched, and complicated by a substantial body of scholastic 'literature', above all in commentaries, paraphrases, and *quaestiones*. Throughout Dante's works, and perhaps above all in the *Commedia*, one can detect the presence of this complex and highly freighted 'Aristotle', one mediated in multiple ways, both textually and orally. Commentators and scholars have long recognised the traces of what we usually refer to as the tradition of medieval Aristotelianism in Dante. Within a few years of his death, in fact, Dante's early commentators had embraced a vision both of him and of his poem as imbued with, and even grounded in, Aristotelian philosophy. In fact, such a perception of the poem constitutes a major critical strand in much of the 14th-century commentary tradition, and it persists well up to the 17th century, returning in modified forms in the late 19th and 20th century.[2] It is a vision that achieves perhaps its ultimate expression in the paean delivered by Dante's great Renaissance commentator, the Florentine Cristoforo Landino, in the prologue to his 1481 print edition and commentary. Here, for Landino, Dante is not only a great "physico" or natural philosopher, but also an indefatigable student of the heavens, of the transformations of the elements and of weather effects, of the human soul and its powers and of the geographical sites of the earth:

> What natural philosopher ever wrote with greater lucidity about all natural motion, according to place or form, whether imperfect or

DOI: 10.4324/9781003412601-2

> perfect, animate or inanimate? Which planetary pathway, conjunction or heavenly orbit has been omitted by him? Which transformation of one element into another, which alteration in the atmosphere of hail, rain, wind, lightning and such like, which mineral composition formed under the earth – which of these things that natural philosophers have demonstrated has this poet failed at least to mention? The soul has four powers, and various offices and properties, yet we have true knowledge of them all in Dante. And who does not know with how much true learning and grace he describes now the universal site of the earth and now some particular region of it.[3]

Of course, there is secure evidence for Dante's assimilation of the *libri naturales* and their capillary presence in his own works, both the vernacular and the Latin ones. At the same time, however, there has been much discussion and debate, especially in the last three decades, about the precise forms and channels through which Dante received Aristotle's writings. There has also been a concomitant discussion regarding the value of characterising him as "Aristotelian", and the precise extent to which Aristotelianism influenced his theories and poetry.[4] Two main areas of concern stand out here, and will inform the account that follows. The first is the question of how such writings and ideas related to natural philosophy reached Dante, given his status as a learned but lay intellectual and one whose conditions in exile, and access to centres of learning and books, were often complex and at times limited. The second is the status and value of the ways Dante uses natural philosophy in his writings, not least given his pronounced tendency both to transform all he assimilates and to fashion new syntheses out of the diverse and even disparate philosophical, literary, and religious currents of his time.

In light of these backgrounds and concerns, this essay attempts to do three main things. First, it provides a synthetic overview of Dante's interest in the *libri naturales* and medieval natural philosophy, above all in the *Vita nova*, the *Convivio*, and the *Commedia*, charting some of the contexts and complexities of his contacts with the *libri* and discussing the more important recent findings and approaches. It then concentrates more narrowly on the *Commedia* and the specific questions that his masterpiece raises with regard to Dante's use of natural philosophy, offering some examples drawn especially from Aristotelian meteorology. And third, we will explore the likely impact and relevance of one major figure, the Dominican theologian and polymath, Albert the Great, responsible for a truly remarkable endeavour to interpret, systematise, and "perfect" the *libri naturales*, and a figure with whom Dante has often been aligned.

2 Aristotle's *Libri naturales* in the *Vita nova* and the *Convivio*

The work of several generations of scholars has helped to establish some relatively firm coordinates for Dante's knowledge of the *libri naturales*.[5] We know that by the time of the drafting of the unfinished *Convivio* (*c*.1303/4–7/8), he had immersed himself in, and embarked on a period of deep study of, Aristotelian natural and ethical philosophy. Aristotle is, of course, already quoted as "il Filosofo" ("the Philosopher") twice in the *Vita nova* (*c*.1292–4) and the very first reference, in *Vita nova* 25, is a relatively technical one. Here, Dante discusses how Love has been personified by poets as if it had a corporeal form and could move around, and, as part of this account, he notes that only a body is "able to move from place to place by its own virtue, according to the Philosopher" ("localmente mobile per sé, secondo lo Filosofo"). The reference seems to betray a precise knowledge of Aristotle's *Physics*: for example, medieval Latin translations of *Physics* IV, 5, 212b8 often read *mobile alicubi est*. However, as Lorenzo dell'Oso has shown in an important study of Dante's intellectual culture, this reference and a further one to Aristotle in *Vita nova* 41.1 (the bat eye simile of *Metaphysics* II, 1, 993b11), do not reveal direct and deep knowledge of Aristotle's works. What is more, no convincing evidence has yet been produced regarding the presence and circulation of the *libri naturales* in Dante's Florence in the 1280s and 1290s.[6] What Dell'Oso shows, instead, is the notable diffusion of *sententiae* and other excerpts of Aristotle in widely diffused compilations such as the *Auctoritates Aristotelis* and in other forms in which philosophical issues were taught to a broader public in Florence, such as the public format of quodlibetal disputation. Such sources and channels of mediation help to explain and contextualise Dante's use of Aristotle in the *Vita nova*.[7]

By the time we reach the early years of exile, that is, the period from 1303 to 1308, something new and extraordinary has happened. In the *Convivio*, that is, Dante's own self-commentary on three of his own earlier poems – a work replete with encyclopaedic and other learning – we see direct quotation of Aristotle and use of the *libri naturales* on a major scale and for a very wide range of topics. A number of eminent historians of ideas, including most recently Gianfranco Fioravanti, have shown just how deep and direct Dante's knowledge of Aristotelian natural philosophy is by the time of the *Convivio*.[8] This is the case for the *Physica*, *Meteorologica*, *De anima*, *De sensu et sensato*, and *De animalibus* (and of course other Aristotelian works, including the *Nicomachean Ethics*). Dante mentions these books by name, and often with reference to specific books, and he provides a good number of vernacularised quotations and paraphrases of their contents. What is notable, too, is the precision and extent of the use of some works, above all the *Physica*, *De coelo et mundo*, *Meteorologica*, and *De anima*. Thus, for example, there are many direct mentions (eight for the *Physica*; ten for *De anima*) but many indirect ones. Dante refers directly to Aristotle as his authority almost one hundred times, and at one point speaks of him as the "the master and leader of human reason" (*Conv*. IV, vi, 8: "maestro e duca della ragione umana"). Dante now displays a precise and deep reliance on the medieval tradition of Aristotelian commentary – notably Averroes, Albert the Great, and Thomas Aquinas – for terms, concepts, arguments, and debates. At times, too, it is also clear that when he makes very specific references to other authorities – such as, say, Avicenna, Ptolemy, and Albumasar – these can be explained by direct knowledge of the supplementary discussions and digressions provided by earlier Aristotelian commentators in glossing the Stagirite's works. Direct knowledge of Aristotle as mediated through his commentators is the case – to give just some examples – in his discussion of the number of heavens, in *Convivio* II, iii, 3–6, of the galaxy in II, xiv, 6–7, and of vision in III, ix.[9] The discussion of the galaxy is particularly complex but illustrative:

> What Aristotle said is difficult to know because his teaching differs from one translation to the other. And I believe this was due to the mistake of the translators. In the new translation, he seems to say that it is the result of a collection of vapours beneath the stars found in that part and which attract them constantly. However, this does not seem to be the true explanation. In the old translation, he says that the Galaxy is nothing other than a multitude of fixed stars in that part, ones that are so small that we cannot make them out from down here but that what appears is that brightness that we call the Galaxy. And it may well be that the heavenly sphere in that part is thicker and hence retains and projects that light. This is the view that Aristotle, Avicenna, Ptolemy appear to hold.[10]

The passage – as has long been recognised – shows first Dante's awareness of competing versions of the *Meteorologica*: the work was translated twice into Latin, initially in a word-for-word translation of the first three books based on an Arabic paraphrase by Gerard of Cremona in the late 12th-century and then in a complete translation from Greek in the 1260s by William of Moerbeke.[11] Both translations had extensive manuscript circulation: the *versio vetus* or *translatio vetus* is extant in some 107 manuscripts, and Moerbeke's *versio nova* or *translatio nova* in some 170 manuscripts.[12] However, Dante's account is difficult to reconcile entirely with the two translations as we know them, and this has justifiably prompted scholars to examine the commentary tradition. As a result, it has been shown that Dante's account most likely draws on the works of Averroes and Albert the Great in particular. Averroes' *Short Commentary* in particular refers to two main theories for galaxy formation – either it is the product of ignited exhalation of vapours or else of light from a cluster of small stars – and raises the issue of errors in the translation.[13] At the same time, and as several *Convivio* commentators have indicated, Dante's exposition also seems to follow closely the explanation provided in Albert's *Meteora*, his extended paraphrase on the *Meteorologica*. In a manner that is remarkably close to Dante's own wording, Albert had spoken about the density of the orb retaining light, the galaxy being nothing other than many small, contiguous stars in that part of the orb, and – perhaps most significantly still for indicating the likely dependence – Albert had noted that this view was held by Ptolemy and Avicenna.[14] And yet, even if we work with a confluence of commentary texts in this way, the Dantean account of the galaxy does not entirely conform to the earlier accounts. Most recently, Anna Chisena has suggested ways in which the continued imprint of compendia might still be felt in this passage.[15] For similar reasons, Fioravanti, in his own *Convivio* commentary, has hypothesised the possibility of contact with oral traditions of discussion on the *Meteorologica.*[16] Fioravanti does not present any example, but one might think about late scholastic *Sentence* commentaries which contain several questions that focussed on natural science and cosmology and at times made use of the *Meteorologica.* Take the case of the Franciscan theologian such as Richard of Middleton, for example, whose own commentary on Peter Lombard's *Sentences* is found in a manuscript copy (though its arrival is difficult to date) at the Franciscan friary of Santa Croce in Florence. In the fourteenth distinction of book II of his *Sentences*, Richard deals explicitly with Aristotle's discussion of the galaxy in a question regarding whether heavenly spheres without stars are visible. In so doing, Richard mentions translations of the *Meteorologica* and reports several different theories about the galaxy, including ones about the density of the heaven leading to the light of stars being accumulated in one place.[17]

3 From the *Convivio* to the *Commedia*

In Dante's great poem, we find even richer evidence of the pervasive assimilation of Aristotelian lore related to all aspects of the physical cosmos, from the movement of the elements to the powers of the heavens, from mineral, plant, and animal lore to the generation of the human body. As in the *Convivio*, Aristotle is mentioned in approving ways, with a stress up his role as "maestro", indeed his position as the supreme embodiment of human reason is reaffirmed and even extended when he appears as a character in the first circle of Hell, Limbo. Here Dante presents the Stagirite as the "master of those who know" (*Inf.* IV, 131: "maestro di color che sanno"). However, quite unlike anything we find in the *Convivio*, Aristotle's name appears far less often and his primacy is subtly nuanced as the poem progresses. There is, in fact, only one other direct mention of him (*Purg.* III, 43), and this sees him paired with Plato, with both philosophers having the limits of their knowledge stressed and problematised in relation to the mystery of the Incarnation.

All the same, the natural philosophy of a medieval Aristotelian kind, often enriched by the scholastic exegetical tradition, underpins many aspects of the narrative, informing major elements of its structure, blocs of imagery, and the ways characters, landscapes, and perceptions in the afterlife are presented. The poet's expertise in such natural philosophy provides the basis for his extraordinary ability to remake – often through metaphors and similes – natural effects related to the plant and animal kingdom, weather, seas and rivers, mountains, and volcanoes. These interests provide the raw material – again and again – for his celebrated poetic verisimilitude, specific lexical and stylistic choices, and, as noted, many similes and metaphors. What is more, the *libri naturales*, and medieval traditions of commentary and teaching upon them, provide Dante not only with building blocs for the way his diegesis is constructed, but also strongly inform major passages where he sets out an understanding of human sensation and perception (*De anima*; *De sensu et sensato*: see *Purg*. XVII, 19–33), a doctrine of the formation of the human soul-body union (*De generatione animalium*: see *Purg*. XXV, 37–60), and parts of his cosmological framework (*De caelo et mundo*: see *Par*. II). In all these areas, however, it is important to recognise that, to a greater and more sophisticated degree than in the *Convivio*, medieval Aristotelian elements are blended, enriched, and perfected by other concerns and other traditions, including other philosophical, theological, and literary currents, both in Latin and in the vernacular. Thus, one never finds passages that might somehow be deemed as exclusively Aristotelian, but rather we see material from natural philosophy combined with the imprint of classics, the vulgate Bible, vernacular poetry, and at times operating in dialogue with other philosophical poetry, such as the continuation of the *Roman de la Rose* by Jean de Meun and the canzone "Donna me prega" by Guido Cavalcanti. In *Purgatorio* XXV, for example, significant parts of Statius's discourse on the formation of the male semen as a formative power that becomes an active principle forming the new members of an organism draw on the Aristotelian account of human generation in the *De generatione animalium* II, 1–3. Statius also alludes to Aristotle's emphasis on the divinity of human intellect and its extrinsic relationship to the body (*De generat. animal*. II, 3, 736b). However, the canto also displays a rich syncretic blending of Aristotelianism with biblical and theological motifs: it is a living and enraptured God that breathes in the rational soul (*Purg*. XXV, 70–71) and this soul is created just as the "heat of the sun makes wine" (v. 77). What is more, as Zygmunt Barański has shown in an important essay on Dante's philosophical poetry, *Purgatorio* XXV goes beyond Aristotelian and scientific sources in its close attention to a predecessor vernacular text, Cavalcanti's "Donna me prega", whose rhyme words are echoed throughout the discourse and at the same time Dante overturns Cavalcanti's key concerns.[18] Moreover, Dante's discreet reticence in alluding to the genitalia, in the expression "where it is more beautiful to be silent than to speak" (vv. 43–44: "ov'è più bello / tacer che dire"), is in stark polemic with Meun's openness to plain speech in his direct references to *couilles* ("balls" and "testicles") in the *Roman*.[19]

We can again offer examples from the *Meteorologica* to help to illustrate all the features noted so far. First, the *Commedia* displays technical precision and a wide range of references to terms and concepts derived from this Aristotelian work. We find all the major features of Aristotelian theory such as the role of vapours and exhalations, their main kinds (hot and dry; cool and moist), and the processes by which dry exhalations rise and can be inflamed and those by which they evaporate and are condensed. So too do we find nearly all of the phenomena treated by Aristotle in books 1–3 of the *Meteorologica*. For Book I, we encounter:

- shooting stars (I, 4), in *Par.* XV, 13–18;
- lights and colours in the atmosphere (I, 5), in *Purg.* I, 13–17; XXV, 91–93;
- the Milky Way (I, 8), in *Par.* XIV, 97–100;
- clouds and mist, drizzle and rain (I, 9), in *Purg.* V, 117–120;
- dew and hoar frost (I, 10), in *Purg.* XXI, 47;
- rain, snow, and hail (I, 11–12), in *Purg.* V, 109–111;
- rivers and seas (I, 13) in *Purg.* XXVIII, 121–123.

As for Books II and III, Dante refers in technically informed ways to

- winds (II, 4–6), in *Inf.* III, 133; *Purg.* XXI, 50; *Par.* VIII, 22;
- earthquakes (II, 7–8), in *Inf.* III, 130–131; *Inf.* XII, 12–14; *Purg.* XXI, 55–57;
- thunder and lightning (II, 9), in *Inf.* IV, 2; *Purg.* V, 37–39; XXXII, 109–111; *Par.* XXI, 12, 108, 136–137; XXIII, 40–43, 99; XXXI, 73–74;
- typhoons (II, 9), in *Inf.* V, 31–33;
- thunderbolts (III, 1), in *Par.* VIII, 22–23;
- optical phenomena that appear due to arrays of reflected light in the atmosphere, such as haloes, both solar and lunar (III, 2–3), in *Par.* XXVIII, 22–24 and 32–33;
- the rainbow (III, 4–5), in *Purg.* XXV, 91–93; XXIX, 77–78; *Par.* XII, 11–12; XXXIII, 118;
- mock suns and rods (III, 6), possibly in *Par.* XXVI, 107–108.

Second – and more significant – we find these phenomena being used for a variety of expressive and narrative purposes, often in complex, extended sequences of imagery that distinctively shape the diegesis and leave an imprint on the poem's language and style. Take, for example, *Inferno* III, where the opening of the canto presents us with the poet's description of how the protagonist (vv. 20–31) reacts to the deafening sounds of pain, anger, and cursing emitted by the first group of damned (the passage also consciously modulates and emulates *Aeneid* VI 557–558), by creating an infernal soundscape that whirls and swirls around Dante, making "a tumult that turns forever in the air [...] like sand when a whirlwind blows" (vv. 29–31: "un tumulto / il quale si aggira in quell'aura [...] / come la rena quando turbo spira").[20] The meteorological context is consciously picked up again at the end of the canto when "the tearful earth" (v. 133: "la terra lagrimosa") gives out a wind that emits a flashing crimson light (v. 134: "balenò una luce vermiglia"), causing Dante to faint and fall to the ground overcome by sleep (vv. 135–136).

In both these cases, and especially the second, the context of Aristotle's *Meteorologica* and its medieval commentaries helps us to appreciate aspects of Dante's dramatic use of different species of winds arising from dry vapours imprisoned in the earth, some leading to earthquakes and others to coruscating effects.[21] Or one thinks of the simile in *Purgatorio* V that compares the alacrity of purgating souls desperately speeding towards Dante and Virgil to the luminous traces formed by shooting stars or evening flashes of light in August.[22] Or in even more complex forms, one might consider the way Dante compares the pilgrim's own ecstatic experience of intense supernatural light to the fire being liberated from a cloud as lightning, and how it then expands and shoots downwards to the earth, against nature:

> Come foco di nube si diserra
> per dilatarsi sì che non vi cape,
> fuor di sua natura in giù s'atterra,
>
> la mente mia così, tra quelle dape,
> fatta piu grande, di se stessa uscio,
> che si fesse rimembrare non sape.
> (*Par.* XXIII, 40–45)

> [As fire is unlocked from a cloud, expanding / beyond containment there, and against its / nature downward enearths itself: / so my mind, made larger than itself amid that / feast, went out of itself, and what it became it / cannot remember].[23]

Here – as in *Purgatorio* V – some of the language is subtly interwoven with suggestions from classical texts,[24] but the primary context is natural philosophical and once again Aristotelian meteorology. In Aristotelian theory (*Meteor.* II, 9), dry vapours light up as fire comes out of a cloud as dry exhalation or vapour moves up into the cloud and is ejected and enflamed; that movement of fire downwards against its natural place is an example of violent motion contrary to nature. We might return again to Albert the Great's expansive paraphrase, where he comments precisely on how the dry vapour when mixed with the wet one rises up into the cold regions and condenses, and how such condensation leads to a vehement agitation that causes a fiery inflammation. In fact, Albert speaks of three kinds of thunder accompanying such lightning, one of which sees the cloud violently broken apart.[25]

However, what is most extraordinary about the simile is its blending of natural philosophy with classical and above all religious and mystical discourses. The explosion of lightning out of the cloud is compared to Dante the character's own mind, as it is rent by the intense luminosity emitted all around him in this Heaven. Dante's own mind is caught up between the "dape" (v. 45) the delights of the feast in Heaven. *Daps*, with the sense of a festive or sacrificial food, is a term we find in Virgil (*Aeneides* III, 223–224; *Georgica* IV, 133) but also in Horace's *Epodes* (II, 48) where it refers, in a bucolic context, to bounteous feasts of nature and from here it is found in several Christian writers, including Propertius.[26] As for the notion of mental dilation, the terminology has rightly been situated in relation to mystical writings, above all Richard of St Victor who speaks of mental expansion in terms that Dante appears to adapt and that were widely diffused even in preacher's handbooks.[27] Such a confluence of the mystical, classical and natural philosophical in a single simile offers, then, one peculiarly rich example of how Dante reassembles constellations of texts and traditions in his poem.

A third and final way in which physical science, including meteorology, is used is as part of major doctrinal sections of the *Purgatorio* and the *Paradiso*. The relevant passages are not digressions but carefully articulated sequences, woven into the narrative and structure of the poem, and presenting us directly with the learning processes of Dante-character. We have already noted the importance of sequences such as *Purgatorio* XVIII and XXV and *Paradiso* II. Time and again, in the Dantean afterlife, and especially after *Purgatorio* XV, we are presented with the character's own understanding of a physical cosmos that operates according to the principles of natural philosophy as he attempts to come to terms with the sights, sounds, smells, varying types of perception and experiences available to him. That Dante-character thinks in ways governed by a natural-philosophical worldview is shown by an early exchange with Virgil in *Inferno* VI, 106–111 where, in order to allow the pilgrim to understand the nature of infernal torments after the Final Judgment, his guide instructs him to return *his* "scienzïa" in order to understand whether such punishments will increase when the soul is united with its body. The "science" here is the natural philosophy of a medieval Aristotelian kind, and one strong implication of the passage is that the pilgrim is himself presented to us a sedulous student of such writings. In Virgil's best-known speech in Hell, his 'lesson' on the moral ordering of Hell in *Inferno* XI, Aristotle's books are mentioned explicitly in the canto both as part of the question-and-answer exchange between Dante and Virgil and in the overall presentation of Dante-character's mental equipment. When, for example, Dante-character questions the place of usury within this classification, Virgil explicitly sends him back to the teachings of ancient natural philosophy, as expounded in Aristotle's foundational work, the *Physics*, which is also identified as one of Dante's own 'books'. Indeed, the pilgrim is sent to turn its very pages, and one notes the insistence upon the physical book object ("non dopo molte carte": *Inf.* XI, 102), an emphasis which helps to reinforce – materially and palpably – Dante's familiarity with the *libri naturales*.

If we remain with meteorology, an excellent example of the structuring and anchoring presence in the narrative of a medieval Aristotelian natural philosophical worldview is found during the meeting with Statius in *Purgatorio* XXI. In essence, we learn here that the earthquake and song on Mount Purgatory – two phenomena that had so overwhelmed Dante and Virgil in the previous canto making them "immobile and in suspense" (*Purg.* XX, 139: "immobili e sospesi") and Dante acutely "desiring to know" (v. 146: "desideroso di sapere") – are the result of a soul completing its purgation and being free to ascend to Paradise:

Quei cominciò: "Cosa non è che sanza
ordine senta la religïone
de la montagna, o che sia fuor d'usanza.

Libero è qui da ogne alterazione:
di quel che 'l ciel da sé in sé riceve
esser ci puote, e non d'altro, cagione.

Per che non pioggia, non grando, non neve,
non rugiada, non brina più sù cade
che la scaletta di tre gradi breve;

nuvole spesse non paion né rade,
né coruscar, né figlia di Taumante,
che di là cangia sovente contrade

secco vapor non surge più avante
ch'al sommo d'i tre gradi ch'io parlai,
dov' ha 'l vicario di Pietro le piante

Trema forse più giù poco o assai
ma per vento che 'n terra si nasconda,
non so come, qua sù non tremò mai.

Tremaci quando alcuna anima monda
sentesi, sì che surga o che si mova
per salir sù; e tal grido seconda".
(*Purg*. XXI, 40–60)

[He began: "Nothing that the rule of this / mountain allows can be without order or beyond custom. / This place is free from any alteration: only what / heaven receives into itself by itself can be caused / here, nothing else. / For this reason, non rain, non hail, no snow, no / dew, no frost falls any higher than the little / stairway with but three steps; / clouds do not appear, whether dense or thin; / nor lightning, nor the daughter of Thaumas, who / often changes neighbourhoods down there: / dry vapour does not rise above the highest of the / three steps I mentioned, where Peter's vicar set his feet. / Lower down, there may be tremours great or / little, but up here wind hidden in the earth has / never shaken it, I know not why. / Here the mountain trembles when some soul / feels itself cleansed, so that it rises up or starts to / climb, and that cry seconds it"].[28]

The supernatural explanation for the earthquake provided in the final tercet of this passage is placed into deep relief and precise context by the enumeration of weather effects in the previous four *terzine*, an enumeration that could not be more naturalistic and Aristotelian. The catalogue proceeds from those effects caused by wet exhalation (rain, snow, hail, clouds and mist, dew and frost are all dealt with by Aristotle at *Meteorologica* I, chapters 9–12), to the mention of dry ones such as lightning ("coruscar": *coruscatio* is the term used in Latin translations in commentators such as Albert and Aquinas), and then onto the rainbow (*Meteor.* III, 4–5). But, the Purgatorial earthquake is decidedly not one generated, as Aristotle and his commentators had explained, by a dry exhalation in the earth (*Meteor.* II, 8). What the inclusion of Aristotelian meteorology does above all, then, is to demarcate between natural processes and the heavenly order. The *Paradiso* develops several such extended passages that tell the story of the pilgrim's own education in the afterlife, as he is urged to balance up and discriminate between naturalistic, philosophical methods of enquiry, on the one hand, and truths based on the Bible and revelation and the prophetic revelations of his divinely-willed journey, on the other. In this way, the attention to natural philosophy forms part of Dante's own highly innovative syncretism, as through his own journey as mediated from his first-person perspective as both character and poet, he presents, judges, and often synthesises various kinds and forms of knowledge, and constructs his own teachings.[29]

4 Dante and Albert the Great

There is a longstanding tradition of studies that has examined the "obligations" that Dante owed to Albert the Great. Benvenuto da Imola, one of the most critically alert of all the early 14th-century commentators, was the first to use Albert's scientific and natural philosophical works widely in his Latin commentary on the *Commedia*.[30] After Benvenuto, the single most important authority is Bruno Nardi who repeatedly drew attention to Albert's influence upon Dante, going so far as to argue that Dante's philosophical culture was much closer to Albert's own than it was to Aquinas.[31] Since Nardi's work, a number of important studies have emerged in the last twenty years, and the anniversary year promises to bring more work yet to be fully digested.[32] The attention is thoroughly merited not only because perhaps no figure in 13th-century theological, philosophical and scientific culture has wielded, and continues to exercise such a potent charge as Albert the Great, but also because we can point empirically to Dante's knowledge of his works, above all in the *Convivio*, where Dante refers explicitly to two treatises that acted as a complement to the *De anima*, that is, the *De somno et vigilia* (*Conv.* II, xiii, 5 and IV, xxiv, 6) and the *De intellectu et intelligibili* (*Conv.* III, vii, 3), and also mentions two further treatises on philosophical geography, the *De natura loci* and pseudo-Aristotelian *De proprietatibus elementorum* (both referred to at *Conv.* III, v, 12). At one further point in *Convivio* (IV, xxiii, 13),

Dante refers explicitly to Albert's *Metaura*. In what follows, we sketch a brief coda to this chapter, offering some methodological reflections on pathways for future work on this topic and some of the issues at stake.

One first (and obvious) pathway is a more rigorous and focused study, one carefully contextualised and drawing upon recent work outside Dante Studies in relation to Albert, of the major works in natural philosophy that we know Dante had read. Such an enquiry might begin – obviously enough – with the treatises mentioned in the *Convivio* and the relevant passages there but it should be expanded out. The *Liber de somno et vigilia* opens with a statement that is deeply revealing of Albert's overall programmatic approach to natural philosophy, that is, his concern to render intelligible for the Latins Aristotle's works, not simply through paraphrase but through techniques such as the use of clarifying digressions on obscure or imperfect parts of the corpus, the division of the work, and the coordination of Aristotle's teachings with complementary ideas in Avicenna, Averroes, Alfarabi, Algazali, and Galen.[33] Such a prominent passage in a work incontrovertibly known to Dante is suggestive of all that he could gather from a reading of even a select number of Albert's works not only in terms of information about the natural world but also in terms of a peculiarly acute intellectual voracity and expansiveness, as well a remarkable openness and concordism with regard to Islamic and Jewish sources. Albert's *De somno* pays eloquent testimony to these qualities and is far more than a philosophically informed account of sleep based on the Aristotelian *De insomniis* and *De divinatio per somnum*. The work draws together – to give only some examples – ideas on humans as the image of the world (lib. III, tr. 1, c. 9), on the role of celestial forms being received into the imagination at moments of alienation from the senses (lib. III, tr. 1, c. 4), the multiple forms of vision and prophecy (including divination by dreams), Alkindi's theory of rays (lib. III, tr. 1, cc. 10 and 13), the relationship between the heart and the sun, and the role of emanative fluxus of all things from a first mover (lib. III, tr. 1, cc. 4 and 9).[34] All these doctrines have some place in Dante and require further careful study.

At the same time, it would be worth examining more closely the entire rich series of writings that Albert added to the *Parva naturalia*, including not only the *De intellectu*, but also the *De iuventute et senectute*, the *De memoria et reminiscentia*, the *De morte et vita*, the *De nutrimento et nutribili*, and the *De spiritu et respiratione*. To these, we should certainly add further study of the *De natura loci* and *De causis et proprietatum elementorum*. Nardi drew attention to Dante's account of place in relation to these works,[35] but the enquiry might be expanded to the role of these works in informing Dante's vision of such matters as diversity in the cosmos, its relationship to the heavens, and the ways in which Platonising metaphysical ideas are adapted to Aristotelian physics.[36] The *De causis et proprietatum elementorum*, Albert's continuous commentary on a work attributed to Aristotle (and Albert does not doubt its authenticity), was first translated into Latin from Arabic in the late 12th century. As with many of Albert's works, he inserts interpretations and digressions into his account when expounding the text. Topics of discussion include not only the properties of the elements, but issues such as the place of the sea in relation to the earth, the role of climate on human appearance, the theory of *climata*, and information on seas and rivers. In Albert's expanded and expansive account, he gives independent treatment – through multiple digressions – to the rise and fall of sea, thermal springs, volcanoes, and mountains, and once again provides remarkable attention on the role of celestial causality as well as to the theory of conjunctions.[37] As for the *De natura loci*, this also offers a wealth of material. Albert's philosophical conception of place is articulated into a complex theory of celestial causality, and there is important material on floods, the sun, the theories of Alpetragius, the variability of the fixed stars, and the role of light and the transparent.[38]

Of course, the focused enquiry would merit broadening out to other Aristotelian works on natural philosophy upon which Albert commented and with which we can show Dante's familiarity by the time of the *Convivio*. This includes Albert's paraphrases on *De caelo et mundo* (certainly well known to him), the *De anima*, the *De sensu et sensato*, and the *Metaura* (whose importance we have noted throughout this essay), and most probably the *De animalibus*. It may well include too the *Mineralia* and other short treatises such as the *De natura animae*. Vasoli's *Convivio* commentary remains a very helpful starting point here for further analysis.[39]

A second pathway, one that needs to be developed in tandem with my suggestions for a reinvigorated study of certain works, is a greater concern with the materiality and circulation of Albert's texts and the history of his early reception. Though we have some relevant documentation on the manuscript tradition, relatively little historicist work has been done in this area and almost none in relation to Dante.[40] The approaches developed in the recent work on Dante's intellectual culture in Florence, which has been concentrated on Florence and Florentine studia, might be adapted to consider how Albert's texts were presented in manuscript copies, how they circulated, what texts travelled together, and how they were diffused throughout Italy in different environments such as in University Arts Faculties (above all Bologna and Padua) and in convent schools and other centres of learning. At the same time, the enquiry would be helped by an enriched appreciation of how his texts and ideas were mediated in lay contexts and other vernacular and encyclopaedic texts.

A third and final set of reflections concerns how we approach Albert's possible presence in the *Commedia*. First of all, we probably need to modify Nardi's own view of Albert as the principal philosophical reference for Dante, including in passages that have often been seen as a test case for his influence such as the doctrine of the rational soul in *Purgatorio* XXV. Of course, Albert remains a very important and formative point of reference, but Nardi, writing in polemic and wanting to counterbalance an overemphasis on Aquinas, may well have pushed too hard in such comments. Robert Hollander and Nicola

Fosca have made this very point in their recent commentaries, and Hollander has noted – in glossing Dante's one explicit reference to Albert in *Par.* XI, 98–99 – how few are the likely direct uses of Albert in the poem. Hollander signals – as Benvenuto da Imola also did – a reference to a landslide near Trento (*Inf.* XII, 4–10; cf. Albert, *Meteora*, lib. III, tr. 2, c. 18) and another one to a simile two cantos later (*Inf.* XIV, 31–36) that relates how Alexander the Great ordered his troops to trample down a fiery rain (Albert, *Meteora*, lib. I, tr. 4, c. 8). To these passages, modern commentators have added a scattering of other references.[41] However what might seem like relatively thin pickings should not discourage us from the kind of work suggested earlier in this section, not least if we consider, by way of conclusion, two examples worthy of further study and related to Albert's natural philosophical writings. The first is a set of passages from both *Purgatory* (XX, 13–15; XXX, 109–111, XXXIII, 40–42) and *Paradise* (II, 112–138; VII, 130–141; VIII 127–135; XIII 52–78; XVII, 96; XXVI, 142–148; XXIX, 52–54), in which Dante presents the cosmic system as one in which power is diffused from the motor intelligence of the heavens, with God or divine providence using such intelligence and heavens as instruments to determine natural events and other types of generation. The second is a group of passages from *Inferno* (IX, 28–32; XI, 1–3; XII, 1–9; XVI, 1–3; XVIII, 16–18; XX, 61–78; XXXIV, 106–126) related to the landscapes of Hell, its rocks, underground caverns, water courses, rivers, earthquakes, rocky outgrows and other geomorphic features, as well as the account Virgil gives of the climb out of Hell.[42]

Notes

1 On the translation and diffusion of the *libri naturales*, the standard and still very helpful account is Dod, "Aristoteles latinus".

2 For further discussion and bibliography, see Simon Gilson, "Aristotele fatto volgare" and Gilson, "Declinazioni aristoteliche e platoniche".

3 Cristoforo Landino, *Comento sopra la Comedia*, vol. 1, 266–267: "Qual physico tutti e moti naturali o secondo el luohgo, o imperfecti, o perfecti, o animati, o inanimati con più lucide ragioni mai scripse? Qual corso di stella, qual congiunctione, qual revolutione di cielo è stata da llui pretermessa? Qual tranformatione d'uno in altro elemento, quale alteratione nell'aere o di grandine, piove, venti, saecte o d'altre simili, qual compositione di minere sobto la terra concreate hanno dimostro e physici che questo poeta non habbi almanco accennato? Ha l'anima quattro potentie; ha varii officii et varie proprieta, ma di tutte habbiam vera cognitione apresso di Danthe. Et chi non sa con quanto vera cognitione et leggiadria hora l'universal sito della terra, hora alchuna particolare regione in epsa descrive?". All translations from Italian and Latin are mine unless noted otherwise.

4 See the major re-evaulation offered in Barański, *Dante e i segni*, and the discussion offered in Gilson, "Rimaneggiamenti danteschi di Aristotele"; Gilson, "Dante and Christian Aristotelianism".

5 See especially Moore, *Studies in Dante: First Series*, 92–156 and 305–318; Minio-Paluello, "Dante's Reading of Aristotle"; Boyde, *Dante Philomythes and Philosopher*; Boyde, *Perception and Passion in Dante*; Boyde, *Human Vices and*

Human Worth in Dante. See also the encyclopedia entries for "De Anima", "De Coelo", "De generatione animalium", "De generatione et corruptione", "De Meteoris", "De sensu et sensato", "Fisica", by Enrico Berti in vol. 2 of *Enciclopedia Dantesca*, 325–326, 330–332, 335–336, 336–337, 364–365, 387–388, 933–934.

6 See discussion and documentation in Pegoretti, "Nelle scuole delli religiosi"; Pegoretti, "Lo 'studium' e la biblioteca di Santa Maria Novella".

7 See dell'Oso, *How Dante Became Dante*, 85–89, 95–99, 102–111. See also Robiglio, "Dante e le *Auctoritates Aristotelis*".

8 Successive commentators of the *Convivio*, and perhaps above all the commentary by Cesare Vasoli, have all helped to demonstrate this deep and first-hand acquaintance; see Dante, *Il Convivio*, edited by De Robertis and Vasoli. Further evidence is found in the focussed studies by Minio-Paluello, "Dante's Reading" and now Fioravanti, "Come dice il filosofo".

9 For further discussion, see Gilson, "*Convivio*".

10 *Conv*. II, xiv, 7: "Quello che Aristotile si dicesse non si può bene sapere di ciò, però che la sua sentenza non si truova cotale nell'una translazione come nell'altra. E credo che fosse lo errore delli translatori; ché nella Nova pare dicere che ciò sia uno ragunamento di vapori sotto le stelle di quella parte, che sempre traggono quelli: e questo non pare essere ragione vera. Nella Vecchia dice che la Galassia non è altro che moltitudine di stelle fisse in quella parte, tanto picciole che distinguere di qua giù non le potemo, ma di loro apparisce quello albore, lo quale noi chiamiamo Galassia: e puote essere, ché lo cielo in quella parte è più spesso e però ritiene e ripresenta quello lume. E questa oppinione pare avere, con Aristotile, Avicenna e Tolomeo". Translation mine.

11 For early discussion, see Toynbee, "Some Unacknowledged Obligations of Dante". For recent commentators on this passage, see Vasoli (in Dante, *Il Convivio*, 248–251); Fioravanti (in Dante, *Convivio*, 326–329), and Imbach (in Dante, *Das Gastmahl. Viertes Buch*, 283–287).

12 On Cremona's *translatio vetus*, see Schoonheim, "Introduction" to *Aristotle's "Meteorology"*, xvii–xviii, xx-lx. For Moerbeke's translation, see the critical edition by Vuillemin-Diem of Aristoteles, *Meteorologica, translatio Guillelmi.*

13 Averroes, *In libros Meteorologicorum expositio media*, fol. 411r: "Et ex hoc videtur quod sermo in hoc signo perficitur per hos duos modos, et sic invenimus Aristoteles fecisse, et hoc apparet ex verbis eius in translatione, quae pervenit ad nos [...] Et forte posuit hoc secundum brevitatem, vel forte propter errorem accidentem in translatione." For further discussion of Averroes' treatment, see Lettinck, *Aristotle's "Meteorology"*, 90–95; Schoonheim, "Introduction" to *Aristotle's "Meteorology"*.

14 Albert, *Meteora*, lib. I, tr. 2, ch. 5–6, 21–22: "nihil aliud autem est Galaxia nisi multae stellae parvae contiguae in illo loco orbis [...] Pars illa orbis est spissior, et ideo retinens et repraesentans lumen solis [...] Et haec est sententia Ptolomaei et Avicennae et etiam Aristotilis." These passages are cited by major commentators, beginning with Toynbee and most recently by Fioravanti (see note 11). These points and authorities are not mentioned in the Thomist commentary.

15 Chisena, "Astronomia e meteorologia nel *Convivio*", 106–109.

16 *Convivio*, ed. Fioravanti, 329.

17 See the question "Utrum in caelum fit a nobis visibile in parte non stellata", in Richard of Middleton, *Super quatuor libros Sententiarum*, lib. II, dist. 14, art. 6: 2, at 190a and 191b: "Galaxia est ex partibus primi orbis sicut dicit Philosophus 2 meteu. [*sic*] secundum translationem antiquam [...] quamvis in Galaxia sint multae partes non stellatae, tamen est ibi maior aggregatio luminis, quam in aliis partibus non stellatis, tamen quia caelum densius est in illa parte".

18 See Barański, "Per similitudine di abito scientifico".
19 On this passage, see Montefusco, "*Roman de la Rose*", 140–141.
20 Translation by Durling in Dante, *Inferno*, 55.
21 Aristotle, *Meteor.* I, 13 mentions earthquakes when wind carries dust aloft. See Albert, *Meteora*, lib. III, tr. 1, c. 1: "grossus [vapor] qui residet in visceribus terrae [...] facit terraemotum [...]". For fuller discussion of "tuono" and its implications, see Stabile, "Tuono". Other texts and traditions may be present here; see e.g. Job 3.6: "Noctem illam tenebrosus turbo possideat".
22 Aristotle, *Meteor.* I, 5, 342a34–36: "Apparent autem aliquando nocte serenitate existente consistentia multa phantasmata in caelo" (*translatio nova*).
23 Translation by Durling in Dante, *Paradiso*, 461.
24 See respectively Virgil, *Georgics* I, 365–366; Lucan, *De bello civile* I, 151–157.
25 Albert, *Meteora*, lib. III, tr. 3, cap. 3: "ascendit vapor humidus per naturam, eo quod de natura sit aquae et ab elemento aquae elevatus, pervenit in loco alto ad locum frigidum, et ex frigididate loci spissatur [...] vapor humidus pervenit ad locum frigoris, et incipit comprimi [...] in tali autem compressione vaporis sicci in ventre nubis fit agitatio vehemens vaporis sicci agitatio autem inducit actualem inflammationem [...] cum impetus totus vapor vel maxima eius pars exilit subito erumpens: et rumpens nubem, exilit in aearam et percutit ipsum percussione bona fortissima".
26 For discussion, see Pucci, "Prudentius".
27 For the relevant texts, see Richard of St. Victor, *Beniamin maior* 5.2, col. 170: "*Mentis dilatatio* est quando animi acies latius expanditur et vehementius acuitur, modum tamen humanae industriae nullatenus supergreditur". For discussion, further texts, and earlier bibliography, see Atturo, "Contemplating wonder", 103–106, 110–111, 113–118.
28 Translation by Durling in Dante, *Purgatorio*, 349.
29 On Dante's syncretism, see Barański, "Sulla formazione intellettuale di Dante"; Gilson, "Sincretismo e scolastica in Dante".
30 For a study, see Gilson, "Notes on the Presence of Albert the Great".
31 Nardi, "Raffronti". For discussion of Nardi's contributions in this respect, see Vasoli, "L'immagine di Alberto".
32 See especially the following studies: Vasoli, "Su alcuni 'riscontri' albertini"; Vasoli, "Dante, Alberto Magno", Vasoli, "Fonti albertiane nel *Convivio*"; Fioravanti, "Dante e Alberto Magno". For studies on *Mineralia*, *De animalibus*, the Dionysian commentaries, see Filosa, "Alberto Magno, Dante"; Gilson, "The Anatomy and Physiology of the Human Body", 11–42; Sbacchi, *La presenza di Dionigi*, 1–20. For specific citations, see Falzone, "Sì come dice Alberto". See also the following recent contributions: Loconsole, "Come virtude in pietra"; Malgieri, "Dove 'l'esser di tutto suo contento giace'".
33 Albert the Great, *Liber de somno et vigilia*, lib. I, tr. 1, c. 1, 123a: "Quia vero librum Aristotelis de scientia ista habemus, sequemur eum eo modo quo secuti sumus eum in aliis, facientes digressiones ab ipso ubicumque videbitur aliquid imperfectum vel obscurum dictum, dividentes opus per libros et tractatus et capitula, ut in aliis fecimus. Nos autem omissis operibus quorumdam modernorum, sequemur tantum Peripateticorum sententias et praecipue Avicennae et Averrois et Alfarabii et Algazelis, quorum libros de hac materia vidimus concordantes: tangemus etiam quandoque opinionem Galeni". See also Albert the Great, *Physica*, lib. I, tr. 1, c. 1, 1: "Et addemus etiam alicubi partes librorum imperfectos et alicubi libros intermissos vel omissos, quos vel Aristoteles non fecit vel forte si fecit, ad nos non pervenerunt" (examples here include the *De principiis motus processivi*, *De mineralibus et lapidibus*, *De plantibus*, and *De causis et processu universitatis*).

34 For important recent studies of the *De somno*, see Donati, "Dreams and Divinatory Dreams"; Donati, "Albert the Great as a Commentator"; Ricklin, "Albert the Grand, commentateur".
35 See especially *Conv.* III, iii, 2–5 with discussion in Nardi, *Nel mondo di Dante*, 77.
36 Important recent studies have paid renewed attention to these topics and others; see especially Anzulewicz, "Die platonische Tradition"; Anzulewicz, "Plato and Platonic/Neoplatonic Sources", 599–601. Particulary relevant here is the interest in emanative fluxus in Albert's *De causis et processu universitatis*, lib. 2, tr. 1, c. 1, 59–61.
37 On the *De causis et proprietatum*, in addition to the *editio Coloniensis*, see Albert the Great, *On the Causes and Properties of the Elements*, and above all Rutkin, *Sapientia Astrologica*, esp. 93–115; Palazzo, "The Scientific Significance of Fate"; Palazzo, "Astrology and Politics".
38 On the *De natura loci*, see Gomez, *The Tropics of Empire*, 64–67, 92–93, 235–242, 252–260.
39 Vasoli reproduces lengthy quotations of Albert's commentaries throughout his *Convivio*, see for especially interesting parallels with natural philosophical commentaries, e.g. 186, 216, 235, 242–243, 247–248, 269, 301–305, 317–318, 322, 323, 339–340, 350–351, 371–372, 394–395, 411–412, 440–441, 489–490, 565, 753–755, 758, 767–768, 809–810, 821–822.
40 For census of manuscripts see Fauser, *Codices manuscripti operum Alberti*; Fauser, "Albertus Magnus Handschriften". For some excellent studies of the manuscript tradition, see Donati, "Toward a Critical Edition of Albert the Great's Treatise 'De nutrimento et nutrito'"; Donati, "Albert the Great's Treatise 'De intellectu et intellegibili'". On Albert in the tradition of early Dante commentary, see Ricklin, "Vom 'frate e maestro' zum 'huomo universale'".
41 The commentaries for Hollander and Fosca for *Par.* X, 97–99 can be consulted online at https://dante.dartmouth.edu/. The Dante Dartmouth Project signals the following other references to Albert: Umberto Bosco and Giovanni Reggio (*Par.* IV, 55–57), quoting *De natura animae* II, 7; Nicola Fosca (*Purg.* XVII, 13–17 cf. Albert, *De somno et vigilia*, III, i, 4; *Par.* XIII, 124–26: Albert, *Physica* I, 3).
42 For the importance of Albert's theories of celestial causality, offering an astrologised Aristotelian natural philosophy, see the works by Rutkin and Palazzo cited in note 37.

Bibliography

Primary Sources

Albert the Great. *Liber de somno et vigilia*. Edited by August Borgnet. Paris: Vivès, 1890.

Albert the Great. *Physica*. Edited by Paul Hossfeld, 2 vols. Münster: Aschendorff, 1971.

Albert the Great. *De causis et processu universitatis*. Edited by Winfried Fauser. Münster: Aschendorff, 1993.

Albert the Great. *Meteora*. Edited by Paul Hossfeld. Münster: Aschendorff, 2004.

Albert the Great. *On the Causes and Properties of the Elements*. Translated by Irvene M. Resnick. Milwaukee, WI: Marquette University Press, 2010.

Aristotle. *Meteorologica, translatio Guillelmi de Moerbeka*. Edited by Gudrun Vuillemin-Diem. *Aristoteles Latinus* 25.3.2. Leiden: Brill, 1995.

Aristotle. *Meteorology*. Edited by Pieter L. Schoonheim. *Aristotle's "Meteorology" in the Arabico-Latin Tradition*. Leiden: Brill, 2000.

Averroes. *In libros Meteorologicorum expositio media. Aristotelis opera cum Averrois commentariis*, vol. 5. Venice: Giunta, 1562. [reprint Frankfurt-am-Main: Minerva Press, 1962].

Cristoforo Landino. *Comento sopra la Comedia*. Edited by Paolo Procaccioli, 4 vols. Rome: Salerno Editrice, 2001.

Dante Alighieri. *Il Convivio*. Edited by Domenico De Robertis and Cesare Vasoli. *Opere minori*, vol. 1/1. Milan: Ricciardi, 1988.

Dante Alighieri. *Inferno*. Edited and translated by Robert M. Durling, introduction and notes by Ronald L. Martinez and Robert M. Durling. *The Divine Comedy of Dante Alighieri*, vol. 1. Oxford: Oxford University Press, 1996.

Dante Alighieri. *Purgatorio*. Edited and translated by Robert M. Durling, introduction and notes by Ronald L. Martinez and Robert M. Durling. *The Divine Comedy of Dante Alighieri*, vol. 2. Oxford: Oxford University Press, 2003.

Dante Alighieri. *Das Gastmahl. Viertes Buch. Italienisch-Deutsch*. Translated by Thomas Ricklin, and edited and commented by Ruedi Imbach, with the collaboration of Roland Béhar and Thomas Ricklin. Dante Alighieri. *Philosophische Werke*, vol. 4/IV. Hamburg: Meiner, 2004.

Dante Alighieri. *Paradiso*. Edited and translated by Robert M. Durling, introduction and notes by Ronald L. Martinez and Robert M. Durling. *The Divine Comedy of Dante Alighieri*, vol. 3. Oxford: Oxford University Press, 2010.

Dante Alighieri. *Convivio*. Edited by Gianfranco Fioravanti. *Canzoni*. Edited by Claudio Giunta. *Opere. 2. Convivio. Monarchia. Epistole. Egloge*. Milan: Mondadori, 2014.

Richard of Middleton. *Super quatuor libros Sententiarum*. Brescia: Sabbium, 1591.

Richard of St. Victor. *Beniamin maior*. Edited by Jacques Paul Migne, *Patrologia Latina*, vol. 196, 1–64. Paris: J. P. Migne, 1855.

Secondary Sources

Anzulewicz, Heinrik. "Die platonische Tradition bei Albertus Magnus. Eine Hinführung". In *The Platonic Tradition in the Middle Ages: A Doxographical Approach*, edited by Stephen Gersh and Maarten J. F. M. Hoenen, 207–278. Berlin: De Gruyter, 2002.

Anzulewicz, Heinrik. "Plato and Platonic/Neoplatonic Sources in Albert". In *A Companion to Albert the Great: Theology, Philosophy, and the Sciences*, edited by Irvine Resnick, 599–601. Leiden: Brill, 2012.

Atturo, Valentina. "Contemplating Wonder. 'Ad-miratio' in Richard of St. Victor and Dante". *Dante Studies* 129 (2011): 99–124.

Barański, Zygmunt G. *Dante e i segni. Saggi per una storia intellettuale di Dante Alighieri*. Naples: Liguori, 2000.

Barański, Zygmunt G. "'Per similitudine di abito scientifico': Dante, Cavalcanti and the Sources of Medieval 'Philosophical' Poetry". In *Science and Literature in Italian Culture*, edited by Pierpaolo Antonello and Simon Gilson, 14–52. Oxford: Legenda, 2004.

Barański, Zygmunt G. "Dante and Christian Aristotelianism". In *Reviewing Dante's Theology*, edited by Claire E. Honess and Matthew Treherne, vol. 1, 65–110. Berlin: Peter Lang, 2013.

Barański, Zygmunt G. "Sulla formazione intellettuale di Dante: alcuni problemi di definizione". *Studi e problemi di critica testuale* 90/1 (2015): 31–54.

Berti, Enrico. "De Anima", "De Coelo", "De generatione animalium", "De generatione et corruptione", "De Meteoris", "De sensu et sensato", "Fisica". In *Enciclopedia Dantesca*, edited by Istituto per l'Enciclopedia Italiana, vol. 2, 325–326, 330–332, 335–336, 336–337, 364–365, 387–388, 933–934. Rome: Istituto per l'Enciclopedia Italiana, 1970.

Boyde, Patrick. *Dante Philomythes and Philosopher. Man in the Cosmos*. Cambridge: Cambridge University Press, 1981.

Boyde, Patrick. *Perception and Passion in Dante's 'Comedy'*. Cambridge: Cambridge University Press, 1994.

Boyde, Patrick. *Human Vices and Human Worth in Dante's 'Comedy'*. Cambridge: Cambridge University Press, 2000.

Chisena, Anna Gabriella. "Astronomia e meteorologia nel *Convivio*: il caso dei *Meteorologica*". In *Dante e la molteplicità delle culture nell'Europa medievale*, edited by Giuseppe Ledda, 97–119. Bologna: Bologna University Press, 2022.

dell'Oso, Lorenzo. *How Dante Became Dante: His Intellectual Formation in Florence between "Clerici" and "Laici" (1294–1296)*, unpublished PhD dissertation. University of Notre Dame, 2020.

Dod, Bernard G. "Aristoteles latinus". In *The Cambridge History of Later Medieval Philosophy from the Rediscovery of Aristotle to the Disintegration of Scholasticism*, edited by Norman Kretzmann, Anthony Kenny, and Jan Pinborg, 45–79. Cambridge: Cambridge University Press, 1982.

Donati, Silvia. "Toward a Critical Edition of Albert the Great's Treatise 'De nutrimento et nutrito'. A Study of the Manuscript Tradition". *Documenti e studi sulla tradizione filosofica medievale* 22 (2011): 235–300.

Donati, Silvia. "Albert the Great as a Commentator of Aristotle's *De somno et vigilia*: The Influence of the Arabic Tradition". In *The Parva Naturalia in Greek, Arabic and Latin Aristotelianism: Supplementing the Science of the Soul*, edited by Börje Bydèn and Filip Radovic, 169–210. Cham: Springer, 2018.

Donati, Silvia. "Dreams and Divinatory Dreams in Albert the Great's *Liber de somno et vigilia*". In *Contemplation and Philosophy: Scholastic and Mystical Modes of Medieval Philosophical Thought in Honor of Kent Emery*, edited by Roberto Hofmeister Pich and Andreas Speer, 178–215. Leiden: Brill, 2018.

Donati, Silvia. "Albert the Great's Treatise 'De intellectu et intellegibili': A Study of the Manuscript and Printed Tradition". *Documenti e studi sulla tradizione filosofica medievale* 30 (2019): 159–264.

Falzone, Paolo. "'Sì come dice Alberto in quello libro che fa dello intelletto'. La citazione del *De intellectu et intelligibili* di Alberto Magno in *Convivio* III, VII, 3–4". In *Letteratura e filologia tra Svizzera e Italia. Studi in onore di Guglielmo Gorni*, edited by Maria Antonietta Terzoli, Alberto Asor Rosa, and Giorgio Inglese, vol. 1, 37–56. Rome: Edizioni di Storia e Letteratura, 2010.

Fauser, Winfried. "Albertus Magnus Handschriften". *Bulletin de philosophie médiévale* 24 (1982): 115–129; 25 (1983): 100–120; 26 (1984): 127–151.

Fauser, Winfried. *Codices manuscripti operum Alberti Magni*. Aschendorff: Münster i. W., 1982.

Filosa, Elena. "Alberto Magno, Dante e le pietre preziose: una nota su ambra ed alabastro". *Dante Studies* 122 (2004): 173–180.

Fioravanti, Gianfranco. "Dante e Alberto Magno". In *Il pensiero filosofico e teologico di Dante Alighieri*, edited by Alessandro Ghisalberti, 93–102. Milan: Vita e Pensiero, 2001.

Fioravanti, Gianfranco. "'Come dice il filosofo': Dante e la *littera* di Aristotele". *Italianistica. Rivista di letteratura italiana* 48/1 (2019): 11–50.

Gilson, Simon A. "Notes on the Presence of Albert the Great in Benvenuto da Imola's *Comentum super Dantis Aldigherij Comoediam*". In *Science and Literature in Italian Culture from Dante to Calvino*, edited by Pierpaolo Antonello and Simon A. Gilson, 72–92. Oxford: Legenda, 2004.

Gilson, Simon A. "Rimaneggiamenti danteschi di Aristotele: *gravitas* e *levitas* nella *Commedia*". In *Le culture di Dante. Studi in onore di Robert Hollander. Atti del quarto Seminario dantesco internazionale, University of Notre Dame (Ind.), USA, 25–27 settembre 2003*, edited by Michelangelo Picone, Theodore J. Cachey Jr., and Margherita Mesirca, 151–177. Florence: Franco Cesati Editore, 2004.

Gilson, Simon A. "The Anatomy and Physiology of the Human Body in the *Commedia*". In *Dante and the Human Body: Eight essays*, edited by John C. Barnes and Jennifer Petrie, 11–42. Dublin: Four Courts Press, 2007.

Gilson, Simon A., "'Aristotele fatto volgare' and Dante as 'peripatetico' in Sixteenth-Century Italy". *L'Alighieri* 39 (2012): 31–63.

Gilson, Simon A. "Sincretismo e scolastica in Dante". *Studi e problemi di critica testuale* 90 (2015): 317–339.

Gilson, Simon A. "*Convivio*". In *Dante's Other Works,* edited by Theodore J. Cachey Jr. and Zygmunt G. Barański, 161–165. Notre Dame: University of Notre Dame Press, 2022.

Gilson, Simon A. "Declinazioni aristoteliche e platoniche nella tradizione esegetica dantesca". In *"La compiuta gioia". Dante e la filosofia. Atti del XXV convegno di studi della Società Italiana per lo Studio del Pensiero Medievale (SISPM) (Udine, 9-11 dicembre 2021),* edited by Luca Maria Bianchi, Stefano Pelizzari, and Andrea Tabarroni. Ravenna: Angelo Longo Editore, 2024, 219–244.

Gomez, Wey. *The Tropics of Empire: Why Colombus Sailed to the Indies.* Cambridge, MA: MIT, 2008.

Lettinck, Paul. *Aristotle's "Meteorology" and Its Reception in the Arab World.* Leiden: Brill, 1999.

Loconsole, Mario. "'Come virtude in pietra da corpo nobilissimo celestiale': La filosofia naturale di Alberto Magno nel *Convivio* dantesco". *Medioevo. Rivista di storia della filosofia medievale* 46 (2021): 69–93.

Malgieri, Maria Evelina. "Dove 'l'esser di tutto suo contento giace'. Note sulla creazione mediata in Dante, tra *Liber de causis* e Alberto Magno". *Medioevo. Rivista di storia della filosofia medievale* 46 (2021): 95–120.

Minio-Paluello, Lorenzo. "Dante's Reading of Aristotle". In *The World of Dante*, edited by Cecil Grayson, 61–79. Oxford: Clarendon Press, 1980.

Montefusco, Antonio. "Roman de la Rose". In *The Oxford Handbook of Dante*, edited by Manuele Gragnolati, Elena Lombardi, and Francesca Southerdern, 127–141. Oxford: Oxford University Press, 2021.

Moore, Edward. *Studies in Dante: First Series* [first edition 1896]. Oxford: Clarendon Press, 1969.

Nardi, Bruno. "Raffronti fra alcuni luoghi di Alberto Magno e di Dante". *Giornale storico della letteratura italiana*, 30/240 (1922): 295–303. [reprinted in Bruno Nardi. *Saggi di filosofia dantesca*, 63–72. Florence: La Nuova Italia, 1967].

Palazzo, Alessandro. "The Scientific Significance of Fate and Celestial Influences in Some Mature Works by Albert the Great: *De fato*, *De somno et vigilia*, *De intellectu et intelligibili*, *Mineralia*". In *Per perscrutationem philosophicam: neue*

Perspektiven der mittelalterlichen Forschung. Loris Sturlese zum 60. Geburtstag gewidmet, edited by Alessandra Beccarisi, Ruedi Imbach, and Pasquale Porro, 55–78. Hamburg: Felix Meiner Verlag, 2008.

Palazzo, Alessandro. "Astrology and Politics: The Theory of Great Conjunctions in Albert the Great". *Quaestio* 19 (2019): 173–203.

Pegoretti, Anna. "'Nelle scuole delli religiosi': materiali per Santa Croce nell'età di Dante". *L'Alighieri* 50 (2017): 5–55.

Pegoretti, Anna. "Lo 'studium' e la biblioteca di Santa Maria Novella nel Duecento e nei primi anni del Trecento (con una postilla sul Boezio di Trevet)". In *The Dominicans and the Making of Florentine Cultural Identity*, edited by Johannes Bartuschat, Elisa Brilli, and Delphine Carron, 105–139. Florence: Firenze University Press, 2020.

Pucci, Joseph. "Prudentius, Readings of Horace in the *Cathemerion*". *Latomus* 50/3 (1991): 677–690.

Ricklin, Thomas. "Albert the Grand, commentateur: L'exemple du *De somno et vigilia* III, 1". *Freiburger Zeitschrift fur Philosophie und Theologie* 45 (1998): 131–150.

Ricklin, Thomas. "Vom 'frate e maestro' zum 'huomo universale'. Mit Dantes *Commedia* auf den Spuren des Albertus Magnus". In *Per perscrutationem philosophicam. Neue Perspektiven der mittelalterlichen Forschung. Loris Sturlese zum 60. Geburtstag gewidmet*, edited by Alessandra Beccarisi, Ruedi Imbach, and Pasquale Porro, 315–332. Hamburg: Meiner, 2008.

Robiglio, Andrea. "Dante e le *Auctoritates Aristotelis*". In *Les Auctoritates Aristotelis, leur utilisation et leur influence chez les auteurs médiévaux: État de la question 40 ans après la publication*, edited by Jacqueline Hamesse and José Meirinhos, 187–202. Turnhout: Brepols, 2016.

Rutkin, H. Darrel. *Sapientia Astrologica: Astrology, Magic and Natural Knowledge, ca. 1250–1800*. Cham: Springer, 2019.

Sbacchi, Diego. *La presenza di Dionigi Areopagita nel 'Paradiso' di Dante*. Florence: Olschki, 2006.

Schoonheim, Pieter L. "Introduction" to *Aristotle's 'Meteorology' in the Arabico-Latin Tradition*, xi–xl. Leiden: Brill, 2000.

Stabile, Giorgio. "Tuono". In *Enciclopedia dantesca*, vol. 5, 752–758. Rome: Istituto della Enciclopedia Italiana, 1976.

Toynbee, Paget. "Some Unacknowledged Obligations of Dante to Albertus Magnus". *Romania* 24 (1895), 399–412. [reprinted in Toynbee, *Dante Studies and Researches*, 39–40. London: Menthuen and Co., 1902].

Vasoli, Cesare. "Dante, Alberto Magno e la scienza dei 'peripatetici'". In *Dante e la Scienza. Atti del convegno internazionale, Ravenna, 28–30 maggio 1993*, edited by Patrick Boyde and Vittorio Russo, 55–70. Ravenna: Longo, 1995.

Vasoli, Cesare. "Fonti albertiane nel *Convivio* di Dante". In *Albertus Magnus und der Albertismus. Deutsche philosophische kultur des Mittelalters*, edited by Maarten J. F. Martin Hoenen and Alain de Libera, 33–49. Leiden, New York and Cologne: Brill, 1995.

Vasoli, Cesare. "L'immagine di Alberto Magno in Bruno Nardi". In *Otto Saggi per Dante*, edited by Cesare Vasoli, 117–132. Florence: Le Lettere, 1995.

Vasoli, Cesare. "Su alcuni 'riscontri' albertini nel *Convivio*". In *Otto Saggi per Dante*, edited by Cesare Vasoli, 103–116. Florence: Le Lettere, 1995.

2 Visual Motion Illusions in the Classical Era and in the Middle Ages

Nicholas J. Wade

1 Introduction

Perceptions commence with stimulation of the senses and behaviour can result from this sequence of events. Behaviours provide records of action of the senses. The range of behaviours is broad and we refer to these as observations and we associate them with verbal descriptions. Observations provide the bedrock of perception and records of observation precede records of their verbal descriptions, as in the case of visual art. Verbal descriptions of observations were refined by Greek philosophers, who also introduced theories to account for the characteristics of perception. Relatively little is known about the origins of visual art; examples of marks made on tools and cave walls have been dated to many thousands of years ago. The adoption of experimental methods to record observations is a more recent development. An early example can be found in the work of Claudius Ptolemy on optics.[1] Ptolemy's *Optics* influenced those of Alhazen, who in turn influenced the medieval *Perspectives* of Bacon, Pecham, and Witelo, which provided the basis for Kepler's enquiries in the 17th century.[2]

The senses of all species have become adapted to the demands of their survival and reproduction, and there is a great variety in the ways in which senses have evolved. Humans not only use their senses they muse about them, too. Paradoxically, much of this musing has concerned minor errors of perception (often called illusions) rather than the constancies of what we perceive. For humans, the experience of the world is generally stable, and the ability to perceive it is easily taken for granted. Objects have positions, shapes, and colours that seem to be perceived instantly, and we can reach for them or move to where they are, without any apparent effort. We can recognise small differences between objects and we can categorise them despite small differences. Clearly, there must be some process that gives rise to visual experience, and it is not surprising that throughout history students of the senses have found it fascinating. Questions arise from such considerations and have been asked since antiquity. If what we perceive is what we take to be true or factual about the world, are everyone's experiences the same? What sorts of mistakes do we make in perceiving? Can perceptual experience

DOI: 10.4324/9781003412601-3

be communicated to others? Throughout the ages, writers, like Dante, have tried to find answers to such questions, which can be considered among the most fundamental that can be posed about the human mind.

Perception engages all the senses but the language in which our experiences are expressed tends to reflect the operation of particular senses. It might seem as though the senses work in isolation rather than in concert. The vocabulary of the senses is not evenly distributed either. Vision has the lion's share of words as well as work associated with our perceptual experiences. Moreover, within vision finer distinctions are made. Contour and colour are often considered separate features of the objects to be processed. This results in space and colour being treated as independent aspects of vision, and students tend to pursue one or the other. Motion presents additional difficulties but illusions of motion were recorded in classical times.

For two thousand years, prior to the scientific enlightenment, the interpretation of mental processes in the Western world was dominated by the wisdom of Greek philosophers. They brought some degree of order to the study of mind, although the ideas they expressed were widely divergent. Much the same applied to the study of vision. The principal contrast that existed was between those who sought to observe and try to account for those observations and those who considered that thought was above perception. The observational tradition has grappled with the vagaries of perception – the differences that occur over time and the contrasts that can take place in an instant. The representatives of this approach are taken as Aristotle and his pupil, Theophrastus. The rational tradition searched for constancies beyond the moment of perception. For some, like Plato, these were to be found in ideal forms that were not expressed via perception but by thought. For others, like Euclid, recourse to mathematics and optics was the way in which the vagaries of vision could be transcended.

2 Visual Illusions

Errors in perception were remarked upon before the basic perceptual processes were either described or appreciated. This was so because it was possible to compare observations of the same objects over time and to note any discrepancies between them. The modern definition of illusions applies to differences between the perception of figures and their physical characteristics. Consensus concerning an external reality did not exist in Antiquity, and so attention was directed to those instances in which changes in perception occurred. That is, when the same object appeared to have different properties under different conditions. According to this observational definition of illusions, all that is required is an assumption of object permanence; thereafter, any changes in the appearance of the same object will be classified as illusions.[3]

While emphasising the veridicality of sensing in general, Aristotle did entertain the possibility of errors (illusions) entering into a particular sense.

Illusions are often considered to be a modern preoccupation, based on specific theories of perception, but their origins are ancient and illusions can be investigated with little in the way of theory. As noted above, if there is an assumption of object permanence, then an illusion occurs when the same object appears to have different properties under different circumstances. Aristotle's description of the motion aftereffect was based on a comparison of percepts. The changing perception of objects was the source of Aristotle's interest in phenomena like afterimages, aftereffects, colour contrasts, and diplopia. It was precisely such variation in perception that led to the Platonic distrust of the senses. Ptolemy drew a distinction between subjective and objective aspects of visual phenomena and devoted considerable space to errors of perception. Alhazen adopted a similar analysis of the errors of direct vision although he extended the range of phenomena for which they occur.

In Antiquity, sight was essential to the study of optics, and disorders of sight influenced theories of vision. The eyes were not only windows on the world, they were also the window through which the world was thought to be illuminated. That is, light itself was considered to have its origin in the eye. This view survived for thousands of years throughout which time light and sight were inextricably intertwined.

Most speculations about sight, advanced by Greek thinkers over many centuries, incorporated contact with external objects. Touch was often taken as the most important sense and the one relative to which others could be related. Thus, vision was generally considered to involve some process of contact between the eye and objects, and several means of achieving this contact were advanced. These included various versions of emission (or extramission) theories, in which light originated in the eye and was projected from it. Reception (or intromission) theories, in which light travelled from objects to the eye, were also advanced, as were speculations incorporating aspects of both emission and reception.

A combination of emission and reception was proposed by Plato, although his theory of vision was always subservient to his philosophy of ideal forms. Plato suggested that light was emitted from both the eye and objects, and vision took place externally where these two streams united. One of the difficulties with this theory was that light continued to be emitted from the open eye at night. Aristotle's theory is more in line with modern conceptions of light, and accounts of it can be found in his books *On the Soul* and *On Sense and the Sensible.*[4] His concern was with observation, and the phenomena he experienced directed the interpretations he proposed. Thus, he queried the emission theories by the simple expedient of testing a prediction that would follow from it: if the light was emitted from the eye, then vision should be possible at any time the eyes were open, including night-time. The fact that the prediction was not supported led him to suggest an alternative theory of the nature of light. Similarly, he distinguished between vision and touch by noting that an object in physical contact with the eye could not be seen.

His alternative interpretation was that vision is the result of some movement in the medium separating the eye from the objects perceived. The emphasis on the medium, variously called the transparent or the diaphanous, reflected Aristotle's distinction between light as a substance, and light as a motion of the medium. Such motions could be instantaneous, and they could be perceived by many observers simultaneously. Aristotle's conception of light was not, however, widely adopted.

Euclid followed Plato's lead and defined optics mathematically, thus equating light and sight. For Euclid, vision was restricted to the cone of rays emanating from the eye and meeting the objects within it. The geometrical projections to these objects were lawful, and this lawfulness was applied to vision, too. Thus, Euclid provided not only an account of optical transmission through space, but also a geometrical theory of space perception itself. The perceived dimensions of objects corresponded precisely to the angles they subtended at the eye, and the illumination of those objects had its source in the eye. The supposed emissions from Euclid's eyes were referred to as visual rays, and their properties were conflated with a number of phenomenal features. The visual rays were discrete, and so small objects could fall between them, and remain unseen; that is, there is a limit to the dimensions of objects that can be detected, namely, a threshold for visual acuity. Moreover, those objects seen by rays in the centre of the visual cone will be seen more clearly than those towards the edge; that is, the direct (foveal) vision has better acuity than indirect (peripheral) vision. It followed, as Euclid stated, that nothing could be seen at once in its entirety, implying that the visual rays would move over an object (by moving the eyes) in order to see all its features.

Greek theories of light were transmitted through the Roman period mostly by Graeco-Roman writers, although the transmission was modulated by a growing desire to integrate optical theories with the practicalities of observation. Lucretius made many references to vision in his poem *On the Nature of Things*; he believed that light (*lumen*) was emitted from the sun, and when it struck objects it carried images (*eidola*) of them to the perceiver. Lucretius followed in the line of the Epicureans, but the relative merits of such reception theories were still in conflict with emission theories, as supported by Hero of Alexandria. He divided the science of vision into three parts: optics, dioptrics, and catoptrics.[5] He considered that the velocity of light was infinite, because of the immediate visibility of heavenly bodies upon opening the eyes. We know relatively little about Ptolemy's theory of light, because the first book of his *Optics* has not survived, but it can be partially reconstructed. What is clear is that his approach was more experimental than those of his predecessors and that he introduced measurements of both reflected and refracted light. Ptolemy extended Euclid's geometrical optics by incorporating facts of both physical optics and visual perception, and by studying them experimentally. In particular, he appreciated that light rays should be thought of as continuous rather than discrete in the way Euclid had stated. He proposed that colour was an integral component of light, and he conducted experiments on colour mixing using a rotating colour wheel. He argued that visual size cannot be equated with visual angle, and introduced the concept that perceived size was derived from visual angle and distance; that is, he addressed the issue of perceptual constancy. He was in agreement with Euclid about the variations in visual acuity throughout the visual pyramid (rather than cone). The two pyramids of vision (one for each eye) needed to be integrated and he conducted experiments with a board in order to study this binocular combination.[6] Ptolemy also realised that illusions occur in vision: "For there are some errors that are caused in all the senses and others that are confined to things seen, of which some are visual and others are in the mind".[7] He was one of the first writers to provide a detailed account of illusions. Indeed, he devoted over one-third of Book II of his *Optics* to errors of sight; they were classified, and then considered under the headings of colour, position, size, shape, and movement. In short, Ptolemy initiated a reconciliation between physical and psychological analyses of vision which was amplified by Alhazen.[8]

Little was added to optical theory in the late Roman period, and the Greek texts were retained and copied initially in Byzantium and later in Persia and North Africa. In the 5th century, Plato's distinction between the material, sensual body, and the rational soul had been incorporated into Christian theology by St. Augustine, and it even permeated the nature of light: spiritual light was the internal illuminant of ideal forms, and physical light was considered to be analogous to this.[9] The ideal forms were rarely encountered in perspective projections; they were present in the mind and could be illuminated by divine light. Hence, we find the emergence of distinctions between different forms of light – *lux* and *lumen. Lumen* was external light, as from the sun or fire, whereas *lux* was perceived light.

Translations of Greek works into Arabic reached their peak in the 9th and 10th centuries, and they in turn were translated into Latin from the 12th century. Because of strictures against dissection, Galen's anatomy and physiology of the eye were generally accepted by Islamic scholars, but they did extend knowledge of optics. Al-Kindi summarised the principal theories of optics proposed by Greek philosophers. Vision could follow from intromission, as the atomists like Democritus had argued, by extramission after the manner of Euclid's theory, by some form of Platonic interaction, or via some medium. Al-Kindi rejected three of the four possibilities, adopting a Euclidean extramission theory. His rejection of the others was largely a negation of any form of intromission in the process of vision. The difficulty with theories incorporating intromission was conceived in terms of the contrast between optical projection to a point (the eye) and perceptual constancy: the former underwent many variations that were not evident in the latter. This apparent conflict between perspective and perception was to influence medieval scholars, too. Both Avicenna and Alhazen accepted that the crystalline lens was the receptive organ for vision, although Alhazen did hint at times that the retina was involved, too. However, he adopted a theory of light similar to that of Aristotle, in which the medium was of prime importance. Alhazen's book on optics had virtually no impact on his contemporaries, but it was rediscovered almost two centuries later, and translated into Latin in the 13th century.[10] Like Ptolemy, Alhazen focussed on errors of sight:

> There exist three modes of visual perception: pure sensation, recognition, and inference and discernment at the time of perceiving the object. Therefore, an error that occurs in what sight perceives by pure sensation will be an error in the sensation itself. And an error in what sight perceives by recognition will be an error of recognition. And an error in what sight perceives by inference and discernment at the moment of vision will be an error in inference and discernment or in the premises on which the inference and discernment are based.[11]

The translation into Latin of the book on optics by Alhazen awakened Western scholars like Roger Bacon, Witelo, and John Pecham, to the physics of light, its mathematical treatment, and its application to vision. The medieval *Perspectives* were principally about direct vision, that is visual optics rather than catoptrics or dioptrics. They shared a common assumption that vision should be analysed in terms of a pyramid with its base on external objects and its apex located on the surface of or in the eye. This perspective pyramid carried with it the problems posed by Al-Kindi, namely the conflict between optical projection and visual perception. One consequence of this was to treat perception with great suspicion, while accepting the validity of perspective projections. Thus, through much of the late medieval period considerably more attention was directed to physical than to psychological dimensions of optics.[12]

Robert Grosseteste is not considered to have had access to Alhazen's *Perspective*, and his analysis of light and vision was Platonic, with light emitted from the eye interacting with that reflected from objects. The impact of absorbing the optics of Ptolemy and of Alhazen is clear in the contrast between Grosseteste and Roger Bacon. While Grosseteste followed Plato on the nature of light and vision, for Bacon pyramids of light strike the eye, but the physiological dimension remained Galenic. The crystalline lens was still taken to be the 'seat of vision' and 'species' remained a part of the process.

3 Motion Illusions

Illusions involving motion were amongst the earliest recorded because they were visible in the natural world. Those examined here are visual persistence, induced motion, motion aftereffects, and visual vertigo. In the early 11th century, Alhazen distinguished between three conditions under which motion is seen: "Sight perceives motion in one of three ways: by comparing the moving object with other objects, or with a single object, or with the eye itself".[13] Alhazen himself expanded on this in the context of eye movements:

> Sight may perceive motion in any one of these ways even if the eye moves. This happens when sight senses the varying position of the moving object while sensing that the variation is not due to the eye's motion. There is a difference in condition between the variation in position that happens to that object on account of its own motion and the variation in position that happens to it on account of the eye's motion. Thus, when sight senses the varying position of the moving object, and senses that the variation in the object's position is not due to the eye's motion, it will sense the object's motion. The form of the object may move in the eye as a result of the object's motion, but sight will not perceive the objects motion merely through the motion of its form in the eye. Rather, sight perceives motion only by comparing the moving object to other objects in the way we have shown. For the form of a stationary object may move in the eye while the object is at rest, but sight will not in consequence of this perceive the object to be in motion. For if the eye moves in front of visible objects while they are being contemplated, the form of every one of the objects facing the eye, whether they are stationary or in motion, will move on the surface of the eye as the latter moves. But sight has become accustomed to the motion of the objects' forms on its surface when the objects are stationary, and therefore does not judge the object to be in motion on account of the motion of its form, unless the form of another object occurs in the eye and sight perceives the varying positions of the moving object's form relative to the form of another object, or unless a succession of forms takes place in the eye as a result of the rotary motion.[14]

This is an exceedingly insightful passage. If the term 'form' is replaced by 'retinal image' then it specifies the relationship between perceived motion and stability. Motion over the retina can result in perceived motion but it can also yield stability, provided that there is no relative motion between objects. What makes the analysis even more remarkable is that it was made centuries before the dioptrical properties of the eye had been clarified and before the concept of the retinal image had been established. Moreover, it makes the assumption of object permanence explicit and hints that it is a consequence of learning.

3.1 Visual Persistence

Motion illusions involving visual persistence were amongst the first spatiotemporal phenomena subjected to quantification.[15] The basic procedure was initially described by Ptolemy and in more detail by Alhazen. A rapidly moving flame will be seen in positions it no longer occupies; if it is rotated rapidly then a circle will be seen. Under normal circumstances, the moving luminous objects we observe can be resolved during their motion, either because they move slowly or because they can be pursued by the eyes. Certain naturally occurring events (like comets and lightning) do not satisfy these conditions, and they were commented on in antiquity. Aristotle discussed persisting images in general, and he likened the effects to those of a projectile moving through space. However, the examples that he gave were concerned with afterimages and aftereffects rather than with the visually persisting effects of observing projectiles. Seneca's account of shooting stars is remarkable because it attributed the visible trail to the "slowness of vision" rather than to any attribute of the object itself:

> These so-called stars leap out and fly across and on account of their great speed seem to trail a long flame. Our sight does not discern their passing but believes the entire path is on fire wherever they fly. The speed of their transit is so great that its stages are not observable. Only the movement as a whole is grasped. We understand more where the star has gone than where it is. So, it marks the entire route as though by a continuous flame because the slowness of our vision does not follow the successive instants of its flight but sees at the same instant where it started and where it ended. The same thing happens in lightning.[16]

Ptolemy had used the same principle implicitly in his colour top and he made explicit reference to the visibility of rapidly rotating discs and wheels; there is also a brief mention of a rapidly moving flame:

> This phenomenon also occurs in shooting stars whose light seems elongated because of the speed of their course, according to the extent of perceived elongation of distance and the perceived phenomenon that strikes vision [...] Thus, objects which move slowly appear to move rapidly when they disappear quickly from view, like a fire which moves during an instant.[17]

Alhazen also referred to a rotating top and he developed a variation on Ptolemy's flame theme that was replayed constantly throughout the following centuries: rapid motion of a burning stick results in the visibility of the path through which it passes:

> Again, if the motion is circular and extremely rapid, like the [rotary] motion of a top, sight will not perceive it, but will perceive the moving top, if the motion is very rapid, as if it were motionless [...] For let someone take a stick aflame at one of its ends, and let him quickly move it right and left in a dark night: looking at such a flame will find it extended through the interval along which it moves, which interval will be many times larger than the flame's magnitude. For sight can perceive an object's size or position or motion only after a measurable interval of time.[18]

3.2 *Motion Aftereffect*

In Antiquity there was an appreciation of the thresholds of motion: some objects changed their position but did not appear to move (like the sun) and others moved too quickly to be seen (like shooting stars). However, relatively little attention was paid to other aspects of visual motion until the dioptrics of the eye had been adequately described in the 17th century. The motion phenomena that were remarked upon prior to that time were those in which there was some conflict between the vision of the same objects under different circumstances. Induced motion and the motion aftereffect (MAE) were two such instances. The MAE refers to the modification of motion perception following prolonged observation of a regularly moving stimulus. Typically, the MAE involves the apparent motion of a stationary stimulus in the opposite direction to a previously observed one, but it can also result in a change in the apparent velocity of a moving stimulus. The study of the perceptual consequences of prolonged stimulation has a long history, but prior to the early 19th century it was confined principally to afterimages and colour phenomena and rarely included motion.

In the natural environment objects, like the stones on the bank of a river, could appear stationary at one time and move at another – following fixation on flowing water. In his book *On dreams*, Aristotle described the MAE in just such circumstances: "when persons turn away from looking at objects in motion, e.g., rivers, and especially those which flow very rapidly, they find that the visual stimulations still present themselves, for the things really at rest are then seen moving".[19] There has been some debate about whether Aristotle either did not report on the direction of the MAE or reported its direction incorrectly. Verstraten has compared a number of different translations from the Greek text and several interpretations of this section from Aristotle, and he concluded that while the MAE itself was described, its direction was not adequately specified.[20]

The MAE is a successive phenomenon whereas induced motion occurs simultaneously. Induced motion also displays a mismatch between the perception of the same objects under different conditions: a stationary object appears to move in the opposite direction to those moving around it, like the apparent motion of the moon when clouds pass by it. Almost three centuries after Aristotle, Lucretius described both induced motion and the MAE direction:

> when our spirited horse has stuck fast in the middle of a river, and we have looked down upon the swift waters of the stream, while the horse stands there a force seems to carry his body sideways and pushing it violently against the stream, and, wherever we turn our eyes, all seems to be rushing and flowing in the same way as we are.[21]

Verstraten compared different translations from Latin and concluded that Lucretius was the first known writer to have specified the direction of the MAE.[22] The reports of both Aristotle and Lucretius were in the context of water flowing in a river, and a critical 19th-century description of the phenomenon was following fixation on descending water in a waterfall. Unlike most other visual phenomena described by Aristotle, the MAE did not become a staple entry in subsequent texts on optics. One possible reason for this could have been that it was not described by Ptolemy in his *Optics*. By contrast, induced motion was described by Ptolemy and by Alhazen, and it has a much more continuous history than the MAE.

3.3 Visual Vertigo

Vertigo is a disturbance of balance (giddiness), usually associated with visual motion and eye movements; it can be temporarily induced by a number of actions. Because it is associated with a variety of medical conditions there have been many descriptions of it but few involved speculations about its cause.[23] The dizziness and giddiness caused by alcohol, looking down from heights, as well as a variety of ailments were known in Antiquity. As noted in the pseudo-Aristotelian *Problemata*:

> Why is it that to those who are very drunk everything seems to revolve in a circle, and as soon as the wine takes a hold of them they cannot see objects at a distance? [...] objects near at hand are not seen in their proper places, but appear to revolve in a circle.[24]

One of the fullest early accounts of vertigo was given by Theophrastus:

> (Cases of) dizziness come about when either (i) an alien breath or (ii) a fluidity from residue moves around the head, from certain types of nourishment, for example either (from) wine or (from) some other juice, or thirdly (iii) when someone moves his head (round) in a circle. For the region around the brain is fluid by nature, and whenever some alien excess enters into it, it forces its way in, percolating to the vessels, and drives the fluid that is there in a circle, so that in a way the same happens as when someone deliberately moves his head in a circle. For it makes no difference whether the origin of movement is external or internal. The movement being like some whirlpool, and the fluid not being continuous, but part being behind in the movement and some in advance, what settles strikes against (what is moving) and brings it to a halt, because it is not in equilibrium in the same way as (what is moving), is carried in whatever direction (what is moving) flows, and produces the dizziness, and often throws people to the ground.[25]

There were later descriptions of the visual effects of body rotation. Lucretius described it as a children's game: "The room seems to children to be turning round and the columns revolving when they themselves have ceased to turn, so much so that they can hardly believe all the building is not threatening to fall in upon them".[26] This was to become a critical procedure for investigating vestibular function. Ptolemy related vertigo to prior exposure to visual motion: "A continuous revolution of the visual field results in objects appearing to move. Movement of this kind is produced in visual vertigo".[27] Some of these observations were incorporated into Alhazen's book on optics:

> Error may also occur in regard to motion when [the condition of] the sight falls outside the moderate range. For if someone turns himself quickly round several times, then stops, he sees all perceptible objects about him turning, though they are stationary. This occurs to him because the spirit that happens to be in the eye moves in its own place and turns and undulates when the person turns round; then when the person stops after this rapid turning, the motion will persist for a while in the visual spirit, just as motion persists in a body which has been moved round and just released.[28]

3.4 Induced Motion

Induced motion refers to the apparent visual movement of a stationary object, usually in the opposite direction to the real movement of other objects. In the natural environment, it is typically a consequence of the motion of some large visual frame (like clouds or the sea) relative to smaller enclosed visual objects. The importance of frames of reference in the perception of motion is evident from the description by Aristotle: "to persons sailing past the land seems to move, when it is really the eye that is being moved by something else [the moving ship]".[29] Similar statements are given by Lucretius and also by Virgil in the *Aeneid* and by Lucan.[30] Ptolemy noted the phenomenon with regard to water flowing past a stationary boat:

> When a boat is stationary on a calm river flowing without waves, a passenger who does not look at the distant land (but only at the water) thinks that the boat is moving quickly and that the water is still. [...] Since the water appears stationary, the motion must be seen as coming from the boat. However, if the water, land, and boat are observed at the same time, and if we are aware that the land is stationary, we will see the water in motion.[31]

Ptolemy's account displays an appreciation of the relativity of motion perception with respect to the frames of reference that are in operation.

Induced motion is a contrast effect in the motion domain: if several objects move together and another remains stationary, then the latter will appear to move in the opposite direction. What seems to be a clear description of induced motion was given over two thousand years ago by Euclid in his *Optics*. He wrote: "If, when certain objects are moved, one is obviously not moved, the object that is not moved will seem to move backward".[32] This account is rather puzzling, since it flies in the face of Euclid's geometrical account of vision. That is, Euclid sought to provide a theory of space and motion perception that reduced these to projective geometry; size perception was equated with visual angles and motion perception was associated with optical displacements. One possible explanation is that Euclid was assuming that the eyes followed the moving objects. Support for this interpretation derives from the fact that the quotation above is sandwiched between two others that are concerned with eye movements, and both define the perceived motion in terms of optical geometry. For example:

> If, when several objects move at unequal speed, the eye also moves in the same direction, some objects, moving with the same speed as the eye, will seem to stand still, others, moving more slowly, will seem to move in the opposite direction, and others, moving more quickly, will seem to move ahead.[33]

Therefore, if the eye was following the two moving objects, the stationary one would move in the opposite direction with respect to the eye, and this would, according to Euclid's theory, account for the perceived motion.

There is no such doubt about the description given two centuries later by Lucretius of a naturally occurring instance of induced motion: "when the winds carry scattered clouds across the sky in the night time, then the shining stars seem to glide against the clouds and to pass above them in a very different direction from their true one".[34] A distinction is implied between the real and apparent motion of the stars, presumably because the stars appeared to remain stationary when the clouds were not visible.

Perhaps the most common and compelling instance of induced motion is when moving clouds partially obscure the moon rather than the stars, as Alhazen noted:

> An example of the errors that may occur in inference and discernment at the time of perception is provided by sight's perception of the movement of the moon when a thin, discontinuous cloud of variable form moves fast before the moon's surface. Sight will erroneously see the moon as moving rapidly, and therefore commit an error of inference because the movement can be perceived only by inference at the time of sensation; and the cause of this error will be that the moon's distance has very much exceeded the range of moderateness. For when a transparent body moves in front of objects placed on the ground near the eye they are not perceived to be in motion, provided the remaining conditions of veridical perception for these objects are within the range of moderateness. This may be experienced in bodies of the surface of the earth, namely those immersed in clear running water. For, observing an object immersed in clear running water, sight will not perceive that

> object to be in motion, though it perceives the moon as moving behind a cloud. [...] Sight therefore perceives the moon as moving behind the swiftly moving cloud because it correlates a part of the moon with successive parts of the cloud. The reason why sight takes this succession to be due to the moon's motion, not the cloud's, is the following: when the cloud covers the sky or a section of it, its motion will not be immediately perceptible; rather, as sight glances at the cloud in this case it will immediately take it to be stationary. It will only perceive the motion of the cloud if, observing the extremity or part of the cloud, it correlates it with a fixed object on the surface of the ground, or with the eye itself, and contemplates it for a while. If, after a sensible period of time, the cloud is found to have changed position, sight will perceive its motion, but such motion will not appear by glancing at the cloud. If the cloud appears motionless and the moon is found to come in line with one part of it after another, the sight will judge the moon to be in motion. This error is due solely to excessive distance, for no such error occurs with objects of this description when they are near the eye.[35]

This observation has been repeated and reported many times in the last millennium, as it provides one of the most readily visible instances of induced motion in the natural environment.

4 Conclusion

Perception has been a continuing concern of poets and philosophers for many centuries. A common thread that has linked the approaches to perception over this timespan has been a concern with the unusual and atypical; that is with illusions. Illusions have been defined in many different ways, but all have involved comparison with some reference. The constancies of perception have not received the same concerted attention. One of the underlying themes in analyses of vision throughout the classical period is a contrast between optics and observation. Essentially, this was a contrast between interpretations based on perspective or perception. Euclid posed the problem and presented a solution in terms of visual projections: visual size was defined by visual angle, and perception corresponded to the projected size of the stimulus. This remarkably simple equivalence did not, however, match the characteristics of everyday perception, and a novel solution was proposed by Ptolemy and amplified by Alhazen. Ptolemy argued that objects of equal visual subtense but at different distances are not judged as the same size, provided that the distances are correctly estimated. Although both these historical interpretations were based on cognitive processes, their importance lies in distinguishing between projection and perception.

Dante would have been exposed to some of these currents of thought, but he certainly described aspects of the world he perceived, particularly illusions.[36]

Notes

1 See Smith, *Ptolemy's Theory of Visual Perception*; Wade, *A Natural History of Vision*; Webster, "Ptolemy's *Optics*". For the Latin medieval version of Ptolemy's optical work see Lejeune's edition: Claudius Ptolemy, *Optica*.
2 Sabra, "Introduction". Howard and Wade, "Ptolemy's Contributions"; Smith, *Ptolemy's Theory of Visual Perception*.
3 See Wade, *Perception and Illusion*.
4 See Gregoric and Fink, "Sense Perception in Aristotle".
5 See Cohen and Drabkin, *A Source Book in Greek Science*.
6 See Crone, "The History of Stereoscopy"; Howard and Wade, "Ptolemy's Contributions"; Smith, *Ptolemy's Theory of Visual Perception*; Webster, "Ptolemy's *Optics*".
7 Claudius Ptolemy, *Optica* II, 56. My translation.
8 See Sabra, "Introduction".
9 Crombie, *Robert Grosseteste*.
10 See Smith, *Alhacen's Theory of Visual Perception*.
11 Ibn al-Haytham, *Optics*, III, 4.1, 258–259.
12 Ronchi, *The Nature of Light*; Meyering, *Historical Roots*.
13 Ibn al-Haytham, *Optics*, II, 3.173, 193.
14 Ibn al-Haytham, *Optics*, II, 3.181, 194.
15 See Wade, "Early History of Illusions".
16 Seneca, *Naturales quaestiones*, I, 75–77.
17 Claudius Ptolemy, *Optica* II, 61 and 77. My translation.
18 Ibn al-Haytham, *Optics*, III, 3.12, 254 and also III, 7.221–222, 348.
19 Aristotle, *On Dreams* II, 459b17–21.
20 Verstraten, "On the Ancient History".
21 Lucretius, *On the Nature of Things* IV, 309.
22 Verstraten, "On the Ancient History".
23 See Brandt, "Man in Motion".
24 The text quoted here is from the spurious *Problems*, included in the series *The Works of Aristotle* by Ross. See Aristotle [pseudo], *Problemata* III, 9, 872a, translated by Forster.
25 Theophrastus, *On Dizziness*, 185.
26 Lucretius, *On the Nature of Things* IV, 307–309.
27 Claudius Ptolemy, *Optica* II, 73. My translation.
28 Ibn al-Haytham, *Optics*, III, 7.261, 360.
29 Aristotle, *On Dreams* II, 460b.
30 See Brigandì, "Il cambio di sistema di riferimento".
31 Claudius Ptolemy, *Optica* II, 78–79. My translation.
32 Euclid, *Optics*, 371.
33 Euclid, *Optics*, 371.
34 Lucretius, *On the Nature of Things* IV, 311.
35 Ibn al-Haytham, *Optics*, III, 4.7–8, 260–261.
36 See Piccolino, "The Visions of Dante", in this volume.

Bibliography

Primary Sources

Aristotle [pseudo]. *Problemata*. Translated by Susanne E. Forster. *The Works of Aristotle*. Edited by William David Ross, vol. 7. Oxford: Clarendon, 1927.

Aristotle. *On Dreams* [*De somniis*]. Translated by John Isaac Beare and William David Ross. *The Works of Aristotle*. Edited by William David Ross, vol. 3. Oxford: Clarendon, 1931.

Claudius Ptolemy. *Optica*. Edited and translated into French by Albert Lejeune. *L'Optique de Claude Ptolémée dans la version Latine d'après l'Arabe de l'Émir Eugène de Sicile*. Louvain: Université de Louvain, 1956.

Euclid, *Optics*. Translated by Harry Edwin Burton. "The *Optics* of Euclid". *Journal of the Optical Society of America* 35/5 (1945): 357–372.

Ibn al-Haytham. *Optics*. Translated by Abdelhamid I Sabra. *The Optics of Ibn Al-Haytham*, vol. 1. *Books I-III. On Direct Vision*. London: The Warburg Institute, 1989.

Lucretius. *On the Nature of Things* [*De Rerum Natura*]. Translated by W. H. D. Rouse. London: Heinemann, 1975.

Seneca. *Naturales quaestiones,* vol. 1: *Books I-III*. Edited and translated by Thomas H. Corcoran. London: Heinemann, 1971.

Theophrastus. *On Dizziness*. Edited and translated by William W. Fortenbaugh, Robert W. Sharples, and Michael G. Sollenberger, 169–249. *Theophrastus of Eresus: On Sweat, On Dizziness and On Fatigue*. Edited and translated by William W. Fortenbaugh, Robert W. Sharples, and Michael G. Sollenberger. Leiden: Brill, 2003.

Secondary Sources

Brandt, Thomas. "Man in Motion. Historical and Clinical Aspects of Vestibular Function. A Review". *Brain* 114 (1991): 2159–2174.

Brigandì, Ottavio. "Il cambio di sistema di riferimento quale aspetto innovativo della riflessione gnoseologica di Dante". In *Letteratura e Scienze*, edited by Alberto Casadei, Francesca Fedi, Annalisa Nacinovich, and Andrea Torre, 1–4. Rome: Associazione degli Italianisti, 2021.

Cohen, Morris Raphael and Drabkin, Israel Edward. *A Source Book in Greek Science*. Cambridge, MA: Harvard University Press, 1958.

Crombie, Alistair C. *Robert Grosseteste and the Origins of Experimental Science, 1100–1700*. Oxford: Oxford University Press, 1953.

Crone, Robert A. "The History of Stereoscopy". *Documenta Ophthalmologica* 81 (1992): 1–16.

Gregoric, Pavel and Fink, Jakob Leth. "Sense Perception in Aristotle and the Aristotelian Tradition". In *Forms of Representation in the Aristotelian Tradition. Volume One: Sense Perception*, edited by Juhana Toivanen, 15–39. Leiden: Brill, 2022.

Howard, Ian P. and Wade, Nicholas J. "Ptolemy's Contributions to the Geometry of Binocular Vision". *Perception* 25 (1996): 1189–1202.

Meyering, Theo C. *Historical Roots of Cognitive Science: The Rise of a Cognitive Theory of Perception from Antiquity to the Nineteenth Century*. Dordrecht: Kluwer Academic Publishers, 1989.

Ronchi, Vasco. *The Nature of Light. An Historical Survey*. Translated by Vinicio Barocas. London: Heinemann, 1970.

Sabra, Abdelhamid I. "Introduction". Ibn al-Haytham, *Optics*. Translated by Abdelhamid I. Sabra. In *The Optics of Ibn Al-Haytham*, vol. 1. *Books I-III. On Direct Vision*, xix–lxxxvii. London: The Warburg Institute, 1989.

Smith, A. Mark. *Ptolemy's Theory of Visual Perception: An English Translation of the Optics with Introduction and Commentary*. Philadelphia, PA: The American Philosophical Society, 1996.

Smith, A. Mark. *Alhacen's Theory of Visual Perception: A Critical Edition, with English Translation, Introduction, and Commentary, of Alhacen's De aspectibus, the Medieval Latin Version of Alhacen's Kitab al-Manazir*. Philadelphia, PA: American Philosophical Society, 2001.

Verstraten, Frans A. J. "On the Ancient History of the Direction of the Motion Aftereffect Direction". *Perception* 25 (1996): 1177–1187.
Wade, Nicholas J. *A Natural History of Vision*. Cambridge: MIT Press, 1998.
Wade, Nicholas J. *Perception and Illusion. Historical Perspectives*. New York: Springer, 2005.
Wade, Nicholas J. "Early History of Illusions". In *Oxford Compendium of Visual Illusions*, edited by Arthur Shapiro and Dejan Todorovic, 3–37. Oxford: Oxford University Press, 2017.
Webster, C. "Ptolemy's *Optics*, Double-vision, and the Technological Afterimage". *Studies in History and Philosophy of Science* 9 (2022): 191–200.

3 Vision as a Tangible and Dynamic Tool in the *Divine Comedy*. An Overview[1]

Marco Piccolino

1 The Divine Comedy as Visual Poem

The Divine Comedy is the narration of an imaginary journey, yet conveyed via direct testimony and personal, therefore predominantly visual, sensorial experience. No wonder that sight is the main sense to which the poet refers in recollecting, and narrating in poetic form, the extraordinary experience of his wanderings in the Underworld, thus following Aristotle (the *maestro di color che sanno* / "master of those who know") – for whom it is sight that, more than other sensations, leads to the acquisition of knowledge. It is therefore not surprising that 'eyes' is the noun with the greatest recurrence in Dante's poem, as underlined in Tibor Wlassics' fundamental study on Dante's optics and vision.[2] The Hungarian scholar rightly places emphasis on the fact that the overabundance in the *Commedia* of the term designating the visual organs par excellence "says little if we consider the thousand expressions that Dante uses to denote the act of seeing and looking".[3]

It is no coincidence, therefore, that, in his poem, Dante immediately announces: "I will recount the other things I sighted" ("dirò de l'altre cose ch'i' v'ho scorte"), thus conveying from the outset the idea that what he is going to tell in verses is the outcome of visual observation.[4]

This emerges at the beginning of the first canto, *Inferno*, which, parenthetically, is the one least marked by visuality, precisely because of the difficulty of vision in darkness that dominates the "cieco mondo" ("blind world") and "loco d'ogne luce muto" ("place mute of all light").[5] The obscurity of Inferno is attested in numerous terms and verses, such as: "selva oscura" ("dark wood)"; "aura sanza tempo tinta, / come la rena quando turbo spira" ("black and timeless air / as sand is swirled in a whirlwind"); "per l'aere nero e per la nebbia folta" ("through that dim air and murky fog"); "aere sanza stelle" ("starless air"), or "aere grosso e scuro" ("dense and murky air"); "oscura costa" ("dark slope"); "fummo del pantano" (marsh-fumes); "tenebre etterne" ("eternal darkness").[6]

The obscurity of Inferno is, however, not so dim as to prevent the pilgrim-poet from discerning the souls of the damned, even in the dark swamp of the Styx, where "l'acqua era buia assai più che persa" ("the water was darker than the deepest purple"), precisely because of the pilgrim's desire and need to see: "And I,

DOI: 10.4324/9781003412601-4

my gaze transfixed, could see / people with angry faces in that bog" (*Inf.* VII, 109–110: "E io, che di mirare stava inteso / vidi genti fangose in quel pantano").

At the end of the first cantica, after the terrible vision of the traitors in the mouths of the three-faced Lucifer ("vidi tre facce a la sua testa"), the farewell to the reader is again characterised by a visual reference, as of an accomplished mission: "But night is rising in the sky. / It is time for us to leave, for we have seen it all" (*Inf.* XXXIV, 68–69: "Ma la notte risurge, e oramai è da partir, ché tutto avem veduto").

Returning to the first canto of *Inferno*, the nature of the visual experience during the journey about to be undertaken is underlined by Virgil, whose unexpected appearance, in an environment once again marked by the absence of light (*Inf.* I, 60: "là dove 'l sol tace") is rendered by two wonderful verses. A visual-acoustic synaesthesia is preceded by the suggestion of an encounter that seems, at the same time, tactile and visual: "before my eyes a figure showed, / faint, in the wide silence" (*Inf.* I, 63: "dinanzi a li occhi mi si fu offerto / chi per lungo silenzio parea fioco"). Of interest, in this respect, is the strong tactile connotation of the sight implied by the recurring verb *ficcare* (to put, thrust, stick, transfix, strain) referred to the eye or to the vision as, for instance, in *Inf.* IV, 11, in "per ficcar lo viso a fondo" ("to strain my eyes toward the bottom").[7] This serves to underline the realism of the visual experience of Dante's afterlife world.

A tactile connotation of vision emerges vividly in *Inferno* XVIII referring to the episode of Thais, the prostitute, expiating her sins in the first *bolgia* of the eighth circle of the Hell, together with pimps and seducers (one of the moments of the *Commedia* where Dante shows his extraordinary ability to create poetry out of a very sordid and vulgar context).

> Appresso ciò lo duca "Fa che pinghe",
> mi disse, "il viso un poco più avante,
> sì che la faccia ben con l'occhio attinghe
>
> di quella sozza e scapigliata fante".
> (*Inf.* XVIII, 127–130)
>
> [And then my leader said to me: "Try to thrust / your face a little farther forward, / to seize a better picture of the features / of that foul, dishevelled wench down there"].[8]

The same tactile and physical dimension of vision is present in other places of the same canto as, for instance, in the following tercet where Dante writes, with reference to the damned souls:

> Mentr'io andava, li occhi miei in uno
> furo scontrati; e io sì tosto dissi:
> "Già di veder costui non son digiuno".
> (*Inf.* XVIII, 40–42)

> [While I went on, my eye stumbled across / one of them, and quickly I brought out: / "Of seeing that man I have not been starving"].[9]

In the first canto of *Inferno*, Virgil 'the guide', addressing the lost and frightened traveller on the verge of becoming a pilgrim on an extraordinary journey, announces the highly visual connotation of the voyage before him:

> vedrai li antichi spiriti dolenti,
> ch'a la seconda morte ciascun grida;
>
> e vederai color che son contenti
> nel foco, perché speran di venire
> quando che sia a le beate genti.
> (*Inf.* I, 116–120)

> [You shall see those ancient souls in pain / as they bewail their second death. / Then you will see the ones who are content / to burn because they hope to come, / whenever it may be, among the blessed].

Virgil's anaphoric insistence here will return as a personal experience, in *Inferno* IV, with the vision of the "spiriti magni" ("the great spirits"), which is announced by visual notations. This is done in tercets strongly characterised by the theme of vision: the flame that illuminates the darkness being a metaphor for the fame that overcomes the darkness of oblivion "I beheld a blaze of light / that overcame a hemisphere of darkness" (*Inf.* IV, 68–69: "vidi un foco / ch'emisperio di tenebre vincia").

> I' vidi Eletra con molti compagni,
> [...]
> Vidi Cammilla e la Pantasilea;
> da l'altra parte vidi 'l re Latino
> [...]
> Vidi quel Bruto che cacciò Tarquino,
> e solo, in parte, vidi 'l Saladino.
> [...]
> Vidi 'l maestro di color che sanno
> [...]
> quivi vid' ïo Socrate e Platone,
> [...]
> e vidi il buono accoglitor del quale,
> Dïascoride dico; e vidi Orfeo,
> (*Inf.* IV, vv. 121–141 *passim*)

> [I saw Electra with many of her line, / [...] I saw Camilla and Penthesilea. / [...] I saw that Brutus who drove out Tarquinius, / [...] And Saladin I saw, alone, apart. / [...] I saw the master of those who know, / [...] I saw

> Socrates and Plato. / I saw Democritus [...] / I saw the skilled collector of the qualities of things / I mean Dioscorides – and I saw Orpheus].

Dante develops the theme of the expiation of the envious in the second *cornice* (ring) of Purgatory (canto XIII), again by playing on the theme of vision. He does it, by starting from the original meaning of the term that denotes the sin "envy" (from the Latin *invidire*, "to see in a malevolent way"), and is marked by the character of darkness and the impossibility of seeing due to the torture to which the penitents are subjected (their eyelashes are circled by an iron wire):[10]

> E come a li orbi non approda il sole,
> così a l'ombre quivi, ond'io parlo ora,
> luce del ciel di sé largir non vole;
>
> ché a tutti un fil di ferro i cigli fóra
> e cusce sì, come a sparvier selvaggio
> si fa però che queto non dimora.
> (*Purg.* XIII, 67–72)

> [And as the sun is of no profit to the blind, / so Heaven's light denies its bounty / to the shades in the place of which I speak, / for iron wire pierces all their eyelids, / stitching them together, as is done / to the untrained falcon because it won't be calmed].

In the *Paradiso*, a cantica *par excellence* dominated by vision and optics in their multifaceted dimensions, Dante invites us to follow, with our gaze, his sight which turns upwards, to contemplate the point where the motions of the celestial spheres intersect each other:

> Leva dunque, lettore, a l'alte rote
> meco la vista, dritto a quella parte
> dove l'un moto e l'altro si percuote;
> (*Par.* X, 7–9)

> [With me, then, reader, raise your eyes / up to the lofty wheels, directly to that part / where the one motion and the other intersect].

The theme of the gaze returns in the following tercet, referring here to the eternal contemplation of creation by the eye of God:

> e lì comincia a vagheggiar ne l'arte
> di quel maestro che dentro a sé l'ama,
> tanto che mai da lei l'occhio non parte.
> (*Par.* X, 10–12)

[and from that point begin to gaze in rapture / at the Master's work. He so loves it in Himself / that never does His eye depart from it].

Further on, the poet invites the reader to imagine, in visual form, what he (Dante-character) sees, and to retain it firmly:

Imagini, chi bene intender cupe
quel ch'i' or vidi – e ritegna l'image,
mentre ch'io dico, come ferma rupe
(*Par*. XIII, 1–3)

[Let him, who would fully understand / what I now saw, imagine -- and let him, while I speak, / hold that image, steady as a rock].

The capacity for visual imagination is thus stimulated in the reader, together with that of the poet-pilgrim, a character already strongly characterised in the first cantica by the desire to see (*Inf*. IX, 107: "ch'avea di riguardar disio"). This desire appears based on an impulse, sometimes uncontrollable, and occasionally even disrespectful towards the damned souls. And, together with imagination, there is, as well, an appeal to a need of maintaining memory, a fundamental theme for the entire narrative structure of the *Commedia*.

In Paradise, when Beatrice invites Saint Peter to subject the poet-pilgrim to a sort of theological examination the "blissed and beautiful woman" addresses the apostle with a highly visual reference:

S'elli ama bene e bene spera e crede,
non t'è occulto, perché 'l viso hai quivi
dov'ogne cosa dipinta si vede;
(*Par*. XXIV, 40–42)

[Whether his love is just, and just his hope and faith, / is not concealed from you because your sight / can reach the place where all things are revealed in a painted form].

Here, the final verse of the tercet gives us the idea of the visual immediacy of knowledge experienced by the blessed. In addition, it contains an allusion to the depictions in church paintings and stained-glass windows or in manuscript illuminations as a means for vehiculating theological issues through the vividness of visual representations.[11]

There are few parts of the poem in which a visual, figurative, or optical, reference is absent. In particular, as a gaze full of curiosity. This is a characteristic that undoubtedly unites Dante as character with Dante as author and denotes an inclination to explore and investigate the world around, sometimes – as we have noted – in an indiscreet way. This tendency is further evidenced in Dante's observation of characters or places, scenes, or choreographies, actually seen or contemplated (in the form of poetic imagination); other times invoking the fiction of dreams or visions, to transfigure – in the form of images – stories or characters, or by resorting to objective figurative representations.[12] Another example is in *Purgatorio* X, where the bas-reliefs on the rocky wall contain scenes inspired by the theme of humility, and, moreover, in *Purgatorio* XII, where others figurations on the floor of the cornice treat the theme of the punishment of pride. This is done according to a procedure undoubtedly inspired by Virgil's representation, in the first book of the *Aeneid*, of the "battles at Troy", painted on the walls of the temple of Juno.

In the *Commedia*, the vision-and-gaze play fulfils another important function, due to the nature – which the poem has – of narrating an event (the pilgrimage into the Underworld), that is supposed to take place precisely at the same time at which it is narrated. Because of this, it necessitates of a tool capable of directing the reader's attention on a particular detail of the scene before the pilgrim's eyes. It allows Dante (as both author and narrator) to isolate and somewhat 'extract' one or more characters from the sometimes-endless multitude of damned, expiating, or blessed souls. This is done with the aim of immortalising his figure in a few verses, in a definitive form, and removed from events of history. Often it is Dante-the-pilgrim who fixes the gaze upon known figures, or upon characters who attract his interest or *curiositas*, due to their condition. Among the most significant examples of this procedure are, in *Inferno* V, the episode of Paolo and Francesca, and in *Inferno* XV, that of Ulysses and Diomedes. On other occasions, it is the soul itself that visually turns towards Dante.

This is, for instance, the case of the meeting, strongly characterised by the theme of visuality, between Dante and Brunetto Latini. Here the pilgrim, after having been eyed by the 'family' of the sodomites, expiating their sins in the fiery sand of Phlegethon, is questioned by his ancient master who takes the helm of his robe:

> E io, quando 'l suo braccio a me distese,
> invidiai li occhi per lo cotto aspetto,
> sì che 'l viso abbrusciato non difese
>
> la conoscenza sua al mio 'ntelletto;
> e chinando la mano a la sua faccia,
> rispuosi: "Siete voi qui, ser Brunetto?"
> (*Inf.* XV, 25–30)

> [And while he held his arm outstretched to me, / I fixed my eyes on his scorched face / until beneath the charred disfigurement / I could discern the features that I knew / and, lowering my hand toward his face, / said: "Are You here, Ser Brunetto?"].

An analogous episode is that of the encounter with Manfredi, where the expiating soul, having been an illustrious and powerful personage in his life, invites Dante to look at him, on the hypothesis that he may have already known him:

> E un di loro incominciò: "Chiunque
> tu se', così andando, volgi 'l viso:
> pon mente se di là mi vedesti unque".
>
> Io mi volsi ver lui e guardail fiso:
> biondo era e bello e di gentile aspetto,
> ma l'un de' cigli un colpo avea diviso.
> (*Purg.* III, 103–108)

> [And one of them began: "Whoever you are, / as you continue walking, turn to look at me, / and think if ever you have seen me in the world". / I turned and fixed my gaze on him. / He was blond, handsome, and of noble aspect, / but a blow had cleft one of his eyebrows].

Authoritative studies have fully examined the symbolic thread of the entire path of asceticism that leads Dante from the uncertain sight in the "dark wood" of the infernal world, up to the beatific vision, in the Empyrean, where the Trinity is an object of visual depictions structured according to an optical-geometric stance: "O eternal Light, abiding in yourself alone" ("O luce etterna che sola in te sidi!"); "lofty Light" ("alto lume"); "having three colours but the same extent" ("tre giri / di tre colori e d'una contenenza"). Here, right at the pinnacle of the contemplative ecstasy of the divinity, in its perennial and complete circularity of knowledge and love, we paradoxically find, in the vision of God, the image of man, "pinta" (painted), represented once again by extensive use of the language of optics and vision. It should be emphasised that this transition from the sight, uncertain and deceptive of the "dark wood", to the vision of Paradise, dominated by a blaze of light and absolute clarity, can only take place through a progressive strengthening of visual capacity, which develops – in parallel with the path of purification and knowledge – somewhat like a gradual 'light adaptation' (to use here – in an anachronistic way – the language of modern sensory physiology).

In the Paradise the vision of the blessed occurs directly, unhindered by obstacles created by "the dim air and murky fog"; and also, without the limits of foreshortening (or distortions or attenuations) imposed by geometric or aerial perspective.[13] In the *Commedia* – let's stress it again – vision serves also to denote knowledge, as noted in the verses, already cited, with the invitation to Saint Peter to examine Dante. In this context, we could also remember that, in *Inf.* VII, 40–41, the moral foolishness of misers and prodigals is rendered by the use of a visual adjective: "All of them had such squinting minds / in their first lives" ("Tutti quanti fuor guerci / sì de la mente in la vita primaia").

In Paradiso vision is so important, to the point of attaining a kind of primacy in the scale of celestial bliss, standing even at a higher grade than love, as the poet tells us in two wonderful tercets:

> Quinci si può veder come si fonda
> l'essere beato ne l'atto che vede,
> non in quel ch'ama, che poscia seconda;
>
> e del vedere è misura mercede,
> che grazia partorisce e buona voglia:
> così di grado in grado si procede.
> (*Par.* XXVIII, 109–114)

> [From this, it may be seen, beatitude itself / is based upon the act of seeing, / not on that of love, which follows after, / and the measure of their sight reveals their worth, /which grace and proper will beget in them. / Such, then, is the process, step by step].

The verses of *Paradiso* XXX describing the boundless contemplation of God by the blessed souls have the effectiveness – admired by Giacomo Leopardi[14] – typical of the creative work of an artist who "not only paints like a master in two strokes [...], but carves and sculpts before the eyes of the reader":

> La vista mia ne l'ampio e ne l'altezza
> non si smarriva, ma tutto prendeva
> il quanto e 'l quale di quella allegrezza.
>
> Presso e lontano, lì, né pon né leva:
> ché dove Dio sanza mezzo governa,
> la legge natural nulla rileva.
> (*Par.* XXX, 118–123)

> [Within that breadth and height, / my sight was not confused but shared / the full extent and quality of that rejoicing. / There, near and far do neither add nor take away, / for where God, unmediated, rules / natural law has no effect].

The ascent of the poet-pilgrim is therefore also – metaphorically – an ascension of a visual nature. Following its development, we are struck by the poet's awareness of how vision (the natural one, of our mundane world) is not a simple passive reception of external signals, but an active and elaborate process, occurring in various phases. In the *Commedia*, these phases are often separated into sequences, which are certainly aspects of the narrative but also indications of a careful reflection by Dante on the perceptive process, that we would be tempted to qualify, with a modern term, as 'physiological'.

2 The 'Cinematographic' Viewpoint in the *Commedia*

As Natalino Sapegno first observed, the appearance of the helmsman angel on the beach of the island of Purgatory (*Purg.* II, 13–51), illustrates Dante's use of a 'cinematographic' procedure for depicting scenes of the afterlife world that gradually appear to the pilgrim's eye, first in the indefinite vision resulting from distance (or from other environmental or subjective conditions that make difficult a distinct perception), and then in the progressive clarification and enrichment of details.

I quote Sapegno's passage at length here because it refers to one of the most beautiful sequences of the second cantica, a sequence characterised by the theme of increasing visual clarity:

> While the two poets pause, uncertain of their path, on the deserted beach of the island, they see a small ship approaching, which glides across the sea not by force of sails or oars, but by a miraculous impulse, under the guidance of an angel who stands straight ahead, stern with wings stretched towards the sky. The vision is portrayed in its movement and its progressive determination: the eye first perceives a light that advances very rapidly (the face of the angel); then, around it, two white masses (the wings), which become visible only later; finally, little by little, as it emerges from the curve of the sea, another white spot (the dress). With a procedure that today would be called cinematographic, the poet distinguishes the successive moments of the show, and at the same time makes one feel the almost instantaneous speed of their succession.[15]

The visual dimension of the sequence is reiterated by Cato, who, addressing Dante, says, with reference to the angel:

> Vedi che sdegna li argomenti umani,
> sì che remo non vuol, né altro velo
> che l'ali sue, tra liti sì lontani.
>
> Vedi come l'ha dritte verso 'l cielo,
> trattando l'aere con l'etterne penne,
> che non si mutan come mortal pelo.
> (*Par.* II, 32–36)

> [Look how he scorns all human instruments / and wants no oar, nor other sail / besides his wings, between such distant shores. / Look how those wings are raised into the sky, / fanning the air with his eternal pinions / which do not change like mortal plumage].

Incidentally, I note here the anaphoric "look" (*vedi*), effective in reiterating the obligatory participation of vision as an intellectual and religious imperative. It is interesting, moreover, to notice that the angel's wings do not resent the effects of the motion through the air. This is an instance of the indifference to natural laws of a world other than the earthly one, a world where – as we have already noticed in reference to Paradise – "natural law has no effect".

The theme of cinematographic vision in the *Commedia*, taken up and extended in a masterly way by Tibor Wlassics in the aforementioned essay, and in subsequent essays of other authors,[16] returns in a recent study[17] in relation to the opening scene of the appearance of the fortifications of the City of Dis:

> Io dico, seguitando, ch'assai prima
> che noi fossimo al piè de l'alta torre,
> li occhi nostri n'andar suso a la cima
>
> per due fiammette che i vedemmo porre,
> e un'altra da lungi render cenno,
> tanto ch'a pena il potea l'occhio tòrre.
> (*Inf.* VIII, 1–6)

> [To continue, let me say that long before / we reached the foot of that high tower / our eyes had noted, at its top / two flaming lights displayed up there / to which another, so far off the eye / could hardly make it out, sent back a signal].

After having evoked the city of Hell which has not yet fully appeared on the visual horizon, the pilgrim-narrator directs our eyes to follow his gaze towards "two flaming lights" moving on top of the "high tower". Immediately afterwards, however, his gaze quickly turns towards another distant and dim light. His focus reflects the experience – common in the Middle Ages – for a citizen, or an armed man (as Dante certainly was), who readily interpreted the significance of light signals between guard posts. Dante (as both author and character) was aware that a signal sent by the sentries of a defensive position would immediately be followed by a response from another post, visually connected to the first. He probably knew as well that the number of torches used by a sentry was correlated to the number of potential enemies approaching (two lights usually indicating a small number).[18]

In this scene, the present-day neuroscience scholar can notice that the difficulty of seeing a distant light depends on the combination of darkness and distance. These factors lead to a decrease in visual contrast, and produce an accentuation of optical distortions, due to the pupillary dilation resulting from dimly lit environments. [19] The visual sagacity of Dante (author) is also evident from his calling attention to the movement of the two flames. It is a moving object that attracts one's gaze, much more than immobile ones, because our eyes are particularly sensitive to movement, especially in the external regions of the visual field.[20]

In this context, it should be considered how the vision of moving images in the visual periphery triggers an innate reflex that leads us to direct our gaze towards the light source, a mechanism of fundamental evolutionary importance for a species, such as the human one, whose vision is characterised by an ability to see details only inside an extremely small area of the visual field. The distinctive vision in this area (corresponding to the fovea) encompasses just about one degree of visual angle, i.e. a circle with a diameter approximately equal to that of the width of our index finger, aimed at the distance of the outstretched arm.

The sequence of the play of light signals approaching the City of Dis illustrates Dante's ability to animate, by the sudden appearance of a new scene, marked by lights and movement, the slow progress of the two pilgrims in "a place mute of all light". At the same time, it reveals an intuitive awareness by the poet of some dynamics of the visual process.

The particular visual force of moving images is even more evident in a sequence from the following canto (*Inf.* IX, 31–36), where what attracts the gaze are not "two flaming lights", but three monstrous figures. Here too, it is a sudden apparition, as unexpected as visually engaging, that distracts the pilgrim from listening to the last words being pronounced by Virgil:[21]:

> E altro disse, ma non l'ho a mente;
> però che l'occhio m'avea tutto tratto
> ver' l'alta torre a la cima rovente,
>
> dove in un punto furon dritte ratto
> tre furïe infernal di sangue tinte,
> che membra feminine avieno e atto.[22]
> (*Inf.* IX, 34–39)
>
> [And he said more, but I do not remember, / for my eyes and thoughts were drawn / to the high tower's blazing peak / where all at once, erect, had risen / three hellish, blood-stained Furies: / they had the limbs and shape of women].

Here too the tower, here too the movement of the gaze from bottom to top, and here too – but in an even more evident way than in the previous canto – the importance of the movement in attracting the visual attention (the three furies stand up with an abrupt motion). And here too – as is obvious for a dynamic form of vision – the movement of the gaze precedes the detailed vision of the object in time and directs attention towards it.

The importance of movement, especially in relation to dynamic scenes with sudden appearances, which attract the spectator and focus his (and our) attention to follow them, is present in other episodes, such as in the lightning-fast manifestation of the two celestial angels:

> Io non vidi, e però dicer non posso,
> come mosser li astor celestïali;
> ma vidi bene e l'uno e l'altro mosso.
> (*Purg.* VIII, 103–105)

> [I did not see and therefore cannot tell / how the celestial falcons started up, / but I could plainly see them both in motion].

To this scene of "celestial falcons", Wlassics refers in underlining the suggestiveness and particular "credibility" of a "spectacle glimpsed only out of the corner of the eye".[23] This is an acute notation by the Hungarian scholar, in which the fact of the particular sensitivity to the movement of the periphery of the visual field re-emerges. Although Dante lacked knowledge of modern sensory physiology, we can say that he displayed, in addition to a great capacity for observation, a particular attention to visual phenomena, a testament to his pictorial sensitivity which is so evident in the *Commedia*.[24]

These and other episodes of the poem can be read by referring to a cinematographic vision, in a kind of 'point of view shot', with the camera that first frames the scene from afar, then focuses on the eyes of the observer, and then moves towards the tower. From here, in the case of the flaming lights, the vision quickly leaps towards the distant light, having first returned for an instant to the pilgrim's eyes, underlining that it is through them that we perceive the dynamism of the scene.

In the case of the Erinyes, the sudden movement towards the high tower is followed by a zooming and focusing action that serves subsequently to reveal the monstrous details of the infernal figures.

3 A Final Remark on Optics and Vision Science in the *Commedia*

For Dante vision was not only a privileged mean useful, in the *Commedia*, for testifying and giving substance to the story of an extraordinary, albeit unreal, existential, and poetic experience. It was also one of the fundamental disciplines of medieval natural philosophy of his age, having been the object of interest and research of important scholars, in particular, the Englishmen Robert Grosseteste, Roger Bacon, and John Peckham, and of the Polish-German Witelo, some of them relying heavily on Arabic texts and especially on Alhazen's *Kitab al-Manazir*. It is therefore not surprising that Dante, in addition to using visual techniques as an expressive and representative means of all his symbolic wanderings in the Underworld, also utilised other dimensions of vision and optics in the *Commedia*. He did this in his desire "to represent – as Eric Auerbach says – the entire earthly-historical world, of which he had knowledge, already subjected to the final judgment of God, and therefore already placed in its proper place in the divine order".[25]

In this context, it is worth remembering how the aspiration to concentrate in an orderly form, in a single work, all the knowledge of the world was in fact a fundamental goal of the medieval scholar, with a millennial history, having its antecedents in the numerous compilations of the classic period. In the *Commedia*, Dante strives to bring together all the knowledge to which he had access, from the biblical, theological, philosophical, and historical sciences, from ethics to art, from classical literary works to mythology, from cosmology to astronomy, from "chorography to architecture"[26] from geography to natural sciences and physics. It is therefore no surprise that he introduces, in his poem, visual science, a blossoming discipline of his age.

It is above all in *Paradiso* and *Purgatorio* that Dante makes use of vision as a science, in particular of what, in a modern sense, we would refer to as "optics". The phenomena of reflection, refraction, and dispersion of light, the formation of rainbows and halos, and other optical phenomena are invoked to give us a somewhat concrete image of a world that should evade, due to its spiritual nature, visual representation. Moreover, they are used to make us participate visually in the harmony of Paradise (this is the case for example of the double rainbow which allows for a physical representation of the double garland of the blessed).

It also provides us with metaphors and physical similes of largely metaphysical 'mysteries': how – for example – it is possible for the spiritual and incorporeal souls of the inhabitants of the afterlife to manifest themselves in corporeal forms. With this aim, Dante first invokes "the air, when it is full of rain" (*Purg.* XXV, 91: "quand'è ben pïorno"), which becomes bright and coloured by dispersing the rays that hit it (*Purg.* XXV, 93: "per l'altrui raggio che 'n sé si reflette, / di diversi color diventa addorno"). In the same way – the poet says – "As soon as space surrounds [the soul] there / the formative force radiates upon it / giving shape and measure as though to living members" (*Purg.* XXV, 88–90: "Tosto che loco lì la circunscrive, la virtù formativa raggia intorno / così e quanto ne le membra vive"), thus generating a visible image, such as a seal would do (*Purg.* XXV, 95–96: "in quella forma ch'è in lui suggella / virtüalmente").[27]

In conclusion, the survey of Dante's commitment to vision and optics in the *Commedia* underlines how his eye and his mind were prepared to glimpse visual phenomena appearing in the world around him and take profit of them for his poetic creation. The following chapter of the present book addresses a simile derived from a personal observation of a visual illusion, that the Florentine poet seems the first to have described.

Notes

1 This chapter is dedicated to Anna Maria, who, since a long time, has been my Virgil in my wandering across literary fields, and recently has re-read with me, with interest and competence, the *Commedia*, when I started venturing in the project on vision in Dante. Thanks to Francesca Galli, Simon Gilson, Cecilia Panti and Mirko Tavoni for their kindness in accepting to discuss with me on this theme and in suggesting relevant bibliography. Thanks, moreover, to Livia Iannucci for their assistance in my archival and bibliographic research, and to Giovanni Niccoli for critically reading an earlier version of this manuscript. Finally, a very great thank to Bernice Braid and Paul Witkovsky for revising the English of this chapter.

2 Wlassics, *Dante Narratore*, particularly 113–140 and 165–182.

3 A similar remark on the frequency of words related to vision and light in Dante's writings has been made by Patrick Boyde in his preface to Gilson, *Medieval Optics*, XIII. This mention is highlighted also in Galli, "Visual Perception", in this volume. To Galli's chapter and Gilson's book I refer for detailed bibliography on vision and optics in Dante's works.

4 I have put "sighted" at the place of "saw" of the Hollanders' translation because it renders better the original verb used by Dante.

5 As Panti ("Seeing the Light", in this volume) remarks, the idea that Dante (the pilgrim) is capable to see in the darkness of the Underworld has led to the wrong tenet that Dante (the poet) adhered to an extramissionist theory of vision.

6 The quotations of Dante' *Commedia* are from the Italian edition of Giorgio Petrocchi (Petrocchi, *Commedia*), and from the English edition of Robert and Jean Hollander (Hollander, *Divine Comedy*) both available at the Danteonline site: https://www.danteonline.it/ (where, however, the title of the translation is incorrectly spelt as *Divine Commedy*). Because of their easy identification through that site, I do not indicate the line number for short quotations inside the text of this and the following chapter. From the same site I have also taken the quotations from the *Convivio*, in the Italian edition of Francesca Brambilla Ageno, and in the English translation of Richard H. Lansing.

7 These are only two of the many instances of Dante's use of the verb *ficcare* referred to the act of vision. A few others are the following: *Inf.* XII, 43: "ficca li occhi a valle" ("But fix your eyes below" [down to the valley]); *Inf.* XII, 43: "Ma ficca li occhi per l'aere ben fiso" ("Now fix your sight more steady through the air"); *Par.* VII, 5: "Ficca mo l'occhio per entro l'abisso / de l'etterno consiglio" ("Now fix your eyes deep in the abyss / of the everlasting will of God").

8 I have substituted here the verb "get", used in the Hollanders' translation of the *Inferno*, with "seize", in order to better render the tactile connotation, present in the original Dante's word, *attinghe*.

9 I have revised here, in an important way, the Hollanders' translation in order to make it adhere to Dante's original, trying particularly to retain the physicality of vision implied in both the first and third verse of the tercet. A very different relation, in Dante, between the visual and tactile dimensions of sensibility has been highlighted in a recent study by Francesco Santi (Santi, "L'esperienza mistica").

10 At the beginning of the episode of the envious, the attention of Dante-the-pilgrim towards the expiating souls, is stated by Virgil with a strong visual reference ("Now fix your sight more steady through the air / and you will make out figures seated / in front of us along the rock"). The theme of vision then returns several times in the following lines ("Then, opening my eyes still wider / I looked ahead and now could see / shades wearing cloaks the color of the stone"; "my eyes were overwhelmed by grief").

11 The stronger effectiveness of the visual language of paintings (and of other forms of figurative representations), compared to verbal messages, is one of the reasons,

according to Thomas Aquinas, that led the Church to promote the decoration with images of worship places. In the third book of his commentary *Super sententias magistri Petri Lombardi*, the great theologian writes that "[images] are there for exciting the feeling of devotion, which is more effectively provoked by things seen than by those heard" ("ad excitandum devotionis affectum qui ex visis efficacius incitatur quam ex auditis": *Super sent.*, lib. 3 d.9 q. 1 a. 2 qc. 2 ad 3).

12 On the theme of the dreams in Dante's *Commedia* see Tavoni, "Where Do Visions Come from?" in this book. See also Tavoni, "L'Inferno sognato".

13 See for instance Panti, *Il visibile "per retta linea*, and her chapter in this volume.

14 Leopardi, *Zibaldone*, 2523: "Dante [...] non solo dipinge senza descrivere [...], ma intaglia e scolpisce dinanzi agli occhi del lettore le proprie idee, concetti, immagini, sentimenti".

15 Sapegno, *Purgatorio*, 15 (in this case, as in all Italian passages that are not taken from published editions in English, the translation is mine). Sapegno had already used the term "cinematographic", in commenting on *Inf.* XXI, 22–30. See Sapegno, *Inferno*, 238: "the rhythmic structure of the line adheres to the movement of the image, which advances rapidly emerging from the background towards the foreground, with a sort of cinematographic procedure. The triplets that follow sculpt the figure of the winged devil, in the snappy and nervous, and possibly threatening and cruel plasticity of his attitude". Here are Dante's verses: "Then I turned like a man, intent / on making out what he must run from, / undone by sudden fear, / who does not slow his flight for all his looking back: / just so I caught a glimpse of some dark devil / running toward us up the ledge / Ah, how ferocious were his looks / and fierce his gesturing, / with wings spread wide and nimble feet!" Here it is worth noting the recurrence of a common expression in the *Commedia* "like a man" ("come uom"), which – according to Wlassics, *Dante Narratore*, 141–164 – would be one of the aspects of Dante's anthropomorphism, that is, a procedure that serves to "humanize" the description of the otherworldly world, introducing a virtual witness ("like a man"), and, often, focusing on his gaze.

16 Among them, Oliva, *Forme di rappresentazione.*

17 Gonzo, *La Commedia degli sguardi.*

18 As Sapegno underlines in his commentary on these verses, Dante also refers to this type of signalling in *Inferno* XXII with the expression "signals from the castle" ("cenni di castella"). The custom of reporting "through a fire or two" any news "in the districts where there is war" is already reported by Boccaccio in his commentary on Dante. See Sapegno, *Inferno*, 246.

19 Nowadays, due to rapid means of transport, the experience of a slow approach towards a distant and gradually more discernible object occurs less often than in Dante's time.

20 A common experience, one easy to test, is to move a computer mouse when we cannot detect the relevant pointer on the screen. In a magical way, the invisible pointer reappears as soon as the hand moves. And this happens, within certain limits, even when the pointer is located in a peripheral area of the visual field.

21 A few centuries after Dante, the awareness of the greater visibility of moving objects will return in a literary work, with Shakespeare. In *Troilus and Cressida*, with the aim of inducing Achilles to emerge from inaction and gain the fame and veneration of the Greeks, Ulysses says: "Since things in motion sooner catch the eye / Than what not stirs" (Shakespeare, *Troilus*, Act III, Scene III, 40). Here the term "catch" brings together the double effectiveness of the movement both to attract the eye and direct its focus.

22 An intense and articulated play of gaze also characterises, in *Inferno* X, the Farinata episode, with, here too, the presence of the movement of the visual object is a fundamental dynamic element of the scene.

23 Wlassic, *Percezione*, 168.
24 In his *Vita di Dante* of 1436, Leonardo Bruni writes that Dante "drew excellently". That he practiced drawing also appears from the *Vita nova*. See Santagata, *Dante*, 77. A great visual sensitivity, the fruit in large part of an important artistic education, also characterised Galileo, who had studied at the *Accademia del disegno* in Florence and was a connoisseur of visual art. As did Dante, certainly Galileo had an irrepressible visual *curiositas*. See Piccolino and Wade, *Galileo's visions*.
25 Auerbach, *Sudi su Dante*, 78–79. It is no coincidence that the great Dante scholar writes that "the content of the *Comedy* is a vision", adding that, in the masterpiece, "the fact seen in the vision is the being and the truth" (at 144).
26 A praise of Dante as "a choreographer and architect of the most sublime judgement" is contained in the first of the two lectures, given by Galileo, on the "figure, site and size of Dante's Inferno", delivered at the *Accademia Fiorentina* somewhat around 1597. (Galilei, *Due lezioni all'Accademia Fiorentina*, 32). In these lectures, Galileo defended the geometric and chorographic conception of Dante's Inferno supported maintained by Antonio Manetti in the 15th century, against the different views held by Alessandro Vellutello and developed in Vellutello's famous commentary to the *Commedia* published in 1544.
27 With relation to Dante's metaphoric use of optics, see Gilson, *Medieval Optics*, and Panti, "Seeing the Light", in this volume.

Bibliography

Primary Sources

Dante Alighieri. *Convivio*. Translated by Richard H. Lansing. *Dante's 'Il Convivio'*. New York: Garland, 1991.

Dante Alighieri. *Convivio*. Edited by Franca Brambilla Ageno, 2 vos., three tomes. Florence: Le Lettere, 1995.

Dante Alighieri. *La Divina Commedia*. Translated by Robert Hollander and Jean Hollander. *The Divine Comedy, Inferno, Purgatorio, Paradiso*. New York: Doubleday & Co., 2003, 2007.

Dante Alighieri. *La Divina Commedia*. Edited by Giorgio Petrocchi, 4 vols. Florence: Le Lettere, 1994.

Galileo Galilei. *Due lezioni all'Accademia Fiorentina circa la figura, sito e grandezza dell'Inferno di Dante*. Edited by Antonio Favaro and Isidoro Del Lungo. *Le Opere di Galileo Galilei, Edizione Nazionale*, vol. IX, 29–57. Florence: Tipografia G. Barbera, 1899.

Giacomo Leopardi. *Zibaldone di pensieri. Edizione tematica condotta sugli indici leopardiani*. Edited by Fabiana Cacciapuoti. Rome: Donzelli, 2023.

Leonardo Bruni. *La vita di Dante e del Petrarca*. Rome: Archivio Guido Izzi, 1987.

Thomas Aquinas. *Scriptum super sententiis magistri Petri Lombardi*. Edited by Maria Fabianus Moos. Paris: Lethielleux, 1956.

William Shakespeare. *Troilus and Cressida*. Edited by Toby Widdicombe. Indianapolis, IN: Hackett, 1967.

Secondary Sources

Gilson, Simon A. *Medieval Optics and Theories of Light in the Works of Dante*. Lewiston, Queenston and Lampeter: The Edwin Mellen Press, 2000.

Gonzo, Elena. *La Commedia degli sguardi. Potenza evocativa e comunicativa degli occhi nel poema dantesco*. Unpublished MA Dissertation at the University of Padova, 2017. Available at: https://thesis.unipd.it/retrieve/a8e149e0-0308-4153-ab690f240adfa12e/ELENA_GONZO_2016.pdf.

Oliva, Gianni. *Per altre dimore. Forme di rappresentazione e sensibilità medievale in Dante*. Rome: Bulzoni, 1991.

Panti, Cecilia. "Il visibile 'per retta linea': Dante, la *perspettiva* e una proposta di emendamento in *Convivio* III ix 6". In *Dante e il mondo: tra realtà e poesia tra storia e letteratura. Atti del* LVIII *Convegno storico internazionale. Todi 10–12 ottobre 2021*, 143–180. Spoleto: Centro Italiano di Studi sull'Alto Medioevo, 2022.

Piccolino, Marco and Wade, Nicholas James. *Galileo's Visions, Piercing the Spheres of the Heavens by Eye and Mind*. Oxford and New York: Oxford University Press, 2014.

Santagata, Marco. *Dante. Il romanzo della sua vita*. Milan: Mondadori, 2017.

Santi, Francesco. "L'esperienza mistica come esperienza tattile in Dante, rileggendo il canto XXXIII del Paradiso". In *L'Esperienza di Dante. Atti del Convegno internazionale di studi*, edited by Giuseppe Ledda, 85–100. Ravenna: Longo, 2023.

Sapegno, Natalino. *La Divina Commedia*, 3 vols. Florence: Nuova Italia, 1985.

Tavoni, Mirko. "L'Inferno sognato, la telepatia di Virgilio e gli antefatti danteschi della *Commedia* come visione in sogno". In *Dante e la dimensione visionaria tra medioevo e prima età moderna*. Atti del Seminario di studio, Freie Universität Berlin, 23 novembre 2017, 97–119. Ravenna: Longo, 2019.

Wlassics, Tibor. *Dante Narratore. Saggi sullo stile della Commedia*. Florence: Olschki, 1975.

4 Moving Clouds and Bending Towers

The Illusive Motion of the Garisenda in *Inferno* XXXI[1]

Marco Piccolino

1 Visual Illusions, the Moving Cloud, and the Bending of the Garisenda

Among the multifarious aspects of Dante's interest in visual and optical phenomena,[2] an episode narrated in the *Inferno*, never examined in depth so far, can be considered an original contribution by Dante to visual science. This observation is most likely based on a personal experience of the poet, since there is, apparently, no mention of it in the optics literature prior to or contemporary with him. Subsequent references to the same phenomenon seem to occur exclusively in the writings by commentators on the *Commedia*, in relation to the same episode.

The event is narrated in the last part of *Inferno* XXXI, the canto of the giants. In the well surrounding the frozen lake of Cocytus, at the level of the ninth circle, there is the place of punishment for those monstrous earthly creatures who, in their violent pride, made war on the Gods or – as the biblical Nimrod – tried to climb the heavens.

The opening of the canto is dominated by the difficulty of seeing, due to the twilight atmosphere that blurs the way of Dante and Virgil: "Here it was less than night and less than day / I could not see too far ahead" (*Inf.* XXXI, 10–11: "Quiv'era men che notte e men che giorno, sì che 'l viso m'andava innanzi poco").[3] The difficulty makes Dante believe, erroneously, that the shapes glimpsed from afar are "lofty towers".[4] Virgil corrects him, pointing out how distance can produce perceptual deceptions. Visual deceptions are indeed a dominant aspect of the canto and one of the fundamental themes of visual science going back to Aristotle. Dante's guide invites him to advance towards those indeterminate objects to figure out the deception: "When you are nearer, you will understand / how much your eyesight is deceived by distance" (vv. 25–26: "tu vedrai se tu là ti congiungi / quanto 'l senso s'inganna di lontano").

It is worth noting here that the theme of sensory deception was one of the dimensions of the visual science to which Dante had access. In the third book of the *Convivio*, the poet refers to the Aristotelian distinction between, on one side, the proper sensibles (such as colours for sight and sounds for

DOI: 10.4324/9781003412601-5

hearing) for which, according to the Greek philosopher, deception is not possible, and, on the other side, the common sensibles. These last ones are susceptible to error because the perceptual outcome of these sensory clues depends on the interaction of different senses, and therefore involves reasoning.

Again, in the *Convivio* (III, ix, 6–7) regarding an argument that the sky can sometimes appear "bright and clear, that is, always with clarity", and sometimes "dark", he writes:

> Here we should know that properly speaking only color and light are visible, as Aristotle asserts in the second book of *On the Soul* and in *On Sense and Sensibles*. It is true that other things are visible, but not properly speaking, because some other sense perceives them, so that they cannot properly be said to be visible, nor, properly, tangible; such are shape, size, number, movement, and state of rest, which we call common sensibles, things that we perceive by more than one sense. But color and light are, properly speaking, visible because we apprehend them by sight alone and by no other sense.[5]

The difficulty of discerning at a distance is used in *Conv*. IV, xii, 15 as an ethical and religious metaphor to signify the deceptions that the soul encounters in its existential wanderings, open, as it is, to misdirection:

> And just as the pilgrim who walks along a road on which he has never traveled before believes that every house which he sees from afar is an inn, and finding it not so fixes his expectations on the next one, and so moves from house to house until he comes to the inn, so our soul, as soon as it enters upon this new and never travelled road of life, fixes its eyes on the goal of its supreme good, and therefore believes that everything it sees which seems to possess some good in it is that supreme good.[6]

Let us return to *Inferno* XXXI, with the two pilgrims who advance in that Underworld towards increasingly darker places (darker from a moral point of view too). At the master's warning about distance as a possible source of visual deception, and the exhortation to proceed onward along the path, Dante realises, as he advances, that those distant shapes "are not towers, but giants", as Virgil had already anticipated him.

In order to describe the increasing visual clarity with decreasing distance, the Latin poet uses a simile derived from a fairly familiar optical experience. As we get closer, the shapes become more and more distinct, "as, when the mist is lifting" (v. 34: "come quando la nebbia si dissipa"), and "little by little we discern things / hidden in the air made thick by fog" (vv. 35–36: "lo sguardo a poco a poco raffigura / ciò che cela 'l vapor che l'aere stipa").[7]

If, however, the visual error progressively dissipates with increasing proximity, on the other hand, approaching the distant objects contributes to increase Dante's trepidation:

> così forando l'aura grossa e scura,
> più e più appressando ver' la sponda,
> fuggiemi errore e cresciemi paura.
> (*Inf.* XXXI, 37–39)
>
> [so, when my eyes saw through the heavy dark / and I got nearer to the brink, / error left me and fear came in its place].

We note here, incidentally, that *Purgatorio* XXIX records another error of judgement due to the distance and intervening medium, similar to that of the giant tower case. Here, Dante initially sees "seven golden trees" in the place of the seven candelabra which precede the procession of atoned souls. This is an interesting episode for us, because – as already noted by Simon Gilson – there is a clear reminiscence of the Aristotelian sensorial conception in the expression "obietto comun" ("common object"), which clearly recalls the Greek philosopher's "common sensible":[8]:

> Poco più oltre, sette alberi d'oro
> falsava nel parere il lungo tratto
> del mezzo ch'era ancor tra noi e loro;
>
> ma quand'i' fui sì presso di lor fatto,
> che l'obietto comun, che 'l senso inganna,
> non perdea per distanza alcun suo atto,
>
> la virtù ch'a ragion discorso ammanna,
> sì com'elli eran candelabri apprese.
> (*Purg.* XXIX, 43–49)
>
> [A short way on, seven golden trees / seemed to appear, an illusion caused / by the space that separated them from us. / But when I had come close enough, / distance no longer could deceive the senses / nor distort the common object's proper shape, / and the faculty that readies reason <for its matter> not clear knew them as candelabra, which indeed they were].

Notice here that the true apperception of the visual object, once the distance has been shortened, implies the intervention of a rational faculty. This is the faculty that in *Paradiso* XXVI, Dante will denote, in another physiological context, as "stimativa", and that in the *Convivio* (III, ix, 9) he indicates, using a paraphrase, as "la parte del cerebro dinanzi – dov'è la sensibile virtute sì come in principio fontale" ("the front part of the brain where the principal source of the sensitive power resides"). This is also a conception of Aristotelian origin.

Returning to *Inferno* XXXI, the theme of the visual error, due to the distance and the weak twilight light, present from the outset of the canto, anticipates the other visual deception that we find towards the end of the canto, namely that of the Garisenda illusive motion. This is a perceptual illusion that Dante describes in an effective and synthetic way in the two tercets that go from line 136 to line 141. Here, reference is made to Antaeus, the tallest and most imposing of the giants. Unlike his companions, however, Antaeus turns out to be a harmless beast, and – indeed – a helpful creature that Virgil convinces to take him and Dante and deposit them on the surface of the frozen lake, where the lower part of Lucifer's body is positioned.

The Latin poet does it with the arts of rhetoric, starting with a *captatio benevolentiae* which serves to make the giant tamer and more accessible. Antaeus then takes in his arms Virgil, who – in turn – holds Dante tightly to him (v. 135: "sì ch'un fascio era elli e io") and places the two "gently" on the surface of Cocytus.

Despite Virgil's reassuring closeness, for Dante-pilgrim the whole action is dominated by fear, a theme that, together that of sensory deception, dominates the canto from beginning to end.

Dante tells us that he would have preferred a different translocation (v. 41: "ch'i' avrei voluto ir per altra strada").

To give us a concrete idea of the fear that assails him as soon as the giant lowers himself to place them on the ground, the poet uses a simile taken from a visual experience. In the act of bending down, Antaeus appears to Dante, who observes him with apprehension, like the Garisenda tower in Bologna, which, under certain conditions, seems to illusively fall on those who look at it from a close distance:

> Qual pare a riguardar la Carisenda [9]
> sotto 'l chinato, quando un nuvol vada
> sovr'essa sì, ched ella incontro penda;
>
> tal parve Anteo a me che stava a bada
> di vederlo chinare, e fu tal ora
> ch'i' avrei voluto ir per altra strada.
> (vv. 136–141)

> [As when one sees the tower called Garisenda / from underneath its leaning side, and then a cloud / passes over and it seems to lean the more, / thus did Antaeus seem to my fixed gaze / as I watched him bend – that was indeed a time / I wished that I had gone another road].

In a few incisive verses Dante describes a visual phenomenon that he – in all probability – observed in his youth while staying, and perhaps studying, in Bologna, the city of many towers, including the two leaning ones, the Garisenda and the Asinelli.[10]

The optical illusion occurs when the observer looks at the Garisenda from the leaning side ("sotto 'l chinato") and a cloud passes across the sky, moving from above the observer towards the bending side of the tower. Under these circumstances, the observer sees the tower move, as if it were collapsing on him. The pilgrim-poet has a similar sensation when, with a sense of trepidation, he looks at Antaeus in the act of bending ("a me che stava a bada di vederlo chinare") to lay the two him and Virgil on the surface of the lake.

The visual illusion described here by Dante represents a kind of *unicum* in the literature of perceptive science.

This is not, however, because the Italian poet was the first, or the only one, to describe the illusory phenomenon in which stationary objects seem to move when other objects in the visual field move in relation to them. Phenomena of this type are well known and, in fact, belong to common experience. The description of them, and attempts at an explanation, are numerous in classical literature, both Greek and Latin, from Euclid and Ptolemy to Lucretius, and in the Arab scholars, Alhazen in particular, up to the moderns, who have studied their mechanisms with experimental methods. These phenomena belong to the class of illusive experiences that vision scholars indicate with the expression "induced-motion", a fairly well-defined subclass of "apparent motion" events.[11]

The circumstances in which illusions of the induced-motion type occur are various, in particular with regard to the conditions that cause the relative motion of objects present in the visual field, or of the observer himself.

Apart from pathological, or – to a certain extent – physiological conditions (such as nystagmus, vertigo, and the forced displacement of an eye), immobile objects in our visual world can appear to us to be moving when we ourselves are the ones actually moving. This happens, for example, when we stay in a train – or in another type of vehicle – which proceeds with uniform motion and without excessive jolts, that is, in a situation in which – due to the principle of inertia – we do not have a direct and unequivocal sensation of movement (like that which can be obtained from the information on the accelerations of our body provided to us by our proprioceptive system).

In a way symmetrical to that described above, still observers can have the sensation of being in motion, when objects or scenes in their field of vision are moving. In the modern world, this is a common experience for a traveller sitting on a train standing in a station, when another train moves on an immediately nearby platform (our apparent movement is still in the opposite direction to that of the other train).[12]

As already noticed by Ptolemy for a similar type of phenomena, if our gaze includes, in addition to the stationary targeted object (and the immediate moving background), also the larger stationary visual world, then we will relate the motion of the background to this, and – consequently – we will correctly perceive the stationary object as immobile. Another characteristic of movement illusions, to some extent related to this one, is the possibility, within certain limits, of controlling the perception of apparent movement with our will, by concentrating our visual attention either on the moving visual background or on the stationary scene surrounding it.

The Garisenda illusion belongs to a class of visual phenomena differing from the ones considered above inasmuch as the illusive movement of a still object is perceived by a still spectator when the immobile visual target is observed against a moving background, this last one appearing – on turn – motionless. The induced motion of this type represents an effect of visual contrast, in the domain of motion perception, between the figure and the background: if the background (or a large enough visual ensemble) moves rapidly, and a particular object remains stationary, then the latter will appear to move in the opposite direction to that of the background. The basis of this particular type of movement illusion had already been hypothesised by Euclid.

The most common experience of this type of effect is that of the Moon which, in the presence of rapidly moving clouds passing in front of it (or approaching or receding from it), seems to move in the opposite direction to the motion of the clouds.

It is possible that Ptolemy was among those describing this specific type of motion illusion as part of his work that has not survived. Several centuries after him, the Arab scholar Alhazen would fully describe the phenomenon of induced illusory motion, in reference to the Moon, with respect to a visual background dominated by moving clouds. He also gives a clear and reasoned indication of the importance of visual reference as an element of explanation of the phenomenon, using the concept of "visual inference". Alhazen follows Aristotle and his doctrine of the possible deceptions with common sensibles, because, for these types of sensibles, perception is the result of a judgement and not of an immediate apprehension, as happens instead for proper sensibles, and he believes that movement, belonging to the class of common sensibles, can be perceived only on the basis of a process of inference.

In book III of the *Kitab al-Manazir*, which deals with "The causes of error of sight", after having discussed various circumstances of visual fallacies attributable to perceptual errors, Alhazen talks about the deceptions surrounding the vision of the movement. In this context, he describes the illusion of the Moon movement caused by moving clouds:

> An example of the errors that may occur in inference and discernment at the time of perception is provided by sight's perception of the movement of the moon when a thin, discontinuous cloud of variable form moves fast before the moon's surface. Sight will erroneously see the moon as moving rapidly, and therefore commit an error in inference because movement can be perceived only by inference at the time

> of sensation; [...] For [sight] perceives the moving body in line with successive parts of the ground-surface on which it moves, getting farther from one part and nearer to another. Sight therefore perceives the moon as moving behind the swiftly moving cloud because it correlates a part of the moon with successive parts of the cloud.[13]

Alhazen's treatment of movement illusions is so clear, and all-encompassing, that anyone who had read it carefully could quite easily have extended it to other situations as well, and – in particular – to circumstances similar to those of the Garisenda illusion described by Dante.

Returning to the poet and his description of the phenomenon in *Inferno* XXXI, there is little doubt that, with his wide familiarity with the visual science of his time, Dante may have had some knowledge of the motion illusion phenomena described by previous authors. He, therefore, might have possessed the cultural background that could allow him to perceive the illusory movement of the Bolognese tower in relation to the movement of the clouds, thus extending to a terrestrial object what Euclid, Lucretius, and especially Alhazen, had earlier observed and discussed for celestial objects.

The fact that no one before Dante had documented the illusion of a tower or a different building bending due to apparent movement would be just one of the many circumstances in which the poet inserted concrete elements into the *Commedia*, borne out from his great curiosity and perspicacity, and based on his personal experience, as well as his vast culture.[14]

Perhaps more surprising is that no reference to the illusory motion of a tower (or of a tall building) is found even in later writings, extending to modern studies of psychophysics. In this regard, we should take into account that, in modern studies, the phenomenon of induced movement is investigated in a controlled way, commonly using artificial patterns generated or projected on screens (the need to quantify the phenomena and to make observations under controlled conditions would in fact make studies based on the inconstancy and variability of cloud movements inappropriate).

It is strange, however, that no other text seems to speak of the Garisenda, or of another tower, leaning or otherwise, that appears to fall at the rapid passing of a cloud. One might have expected that some travellers who journeyed through Italy between the 18th and 19th centuries on their *Grands tours*, often with scientific curiosities no less lively than the interest in the art and history of this country, would have experienced the Garisenda illusion and commented on it.

Charles de Brosses, a French scholar and president of the Parliament of Lyon, spoke of the Garisenda in his family letters from Italy, and in particular in the letter entitled *Mémoire de Bologne* addressed to Madame de Neuilly on 15 September 1739.[15] De Brosses lingers in describing the Garisenda, together with the nearby – and taller – Asinelli tower. He notes the great slope of the first, saying that "if a lead were thrown from its top, it would fall at a distance of more than nine feet from the foundations" (a rather correct observation, because the overhang of the Garisenda – much steeper than its neighbour – is 10.5 feet (i.e., 3.22 meters)). He also writes about the fear that it must exercise on passers-by who believe that it could collapse on them ("qui croyent qu'elle va leur faire calotte"), but he says nothing about the fact that the sensation of imminent collapse may be accentuated by the passage of clouds.

A few decades after De Brosses, Goethe was in Bologna and did not want to pass up the chance to climb to the top of Garisenda. The German poet (then very interested in science) tended to believe that the slope of the tower was not due to a subsidence of the land but rather was deliberately designed to give prestige – with its extraordinary nature – to the family who had built it.[16] That said, he makes no mention of its illusory bending due to the movement of the clouds. It is strange that, in the notes taken on 20 October 1787, two days after the visit to the tower, Goethe refers to Antaeus. However, he does not relate it to the episode of Dante and Virgil's descent towards the bottom of Hell, but to metaphorically express how, as he approached the mountains surrounding the city, his passion for geology was resurrected. This was a passion that had remained with him since the days when he was in charge of supervising the mines of Thuringia: "I am like Antaeus – he writes – who felt invigorated as soon as he set foot on the earth, his mother".[17]

2 The Garisenda in the Comments to the *Commedia*: The Reasons and the Interest of a 'Lack of Interest'

In addition to my scientific training in the field of visual physiology, my personal interest in the Garisenda illusion described by Dante came precisely from the little attention given to it by Dante scholars and, in particular, by the commentators of the *Commedia*. And, also, from the inaccuracies I found among the notes to *Inferno* XXXI in some comments of the school editions of Dante's masterpiece.

Stimulated by Dante's beautiful verses, and above all by their sculptural and incisive brevity, I spent a long period studying the phenomenon of the movement of terrestrial buildings as the clouds move, above, or near them.

The main difficulty in this type of research depends on the fact that, since it involves a study carried out in the field, and not in a laboratory, one is at the mercy of uncontrollable events such as the volatile and inconstant presence of moving clouds in a relatively clear sky. It can therefore happen, even to a scholar extremely interested in the phenomenon who goes to Bologna to directly observe it in conditions closest to that of Dante, that he remains deluded because of ever-changing atmospheric events: precisely what happened to Alfred Bassermann towards the end of the 19th century, because, during his stay in Bologna, "the sky remained inappropriately cloudless".[18]

Faced with the 'deceptive' experience of the German scholar, resulting from the unavailability of moving clouds in the Bologna sky, I am amazed by a passage found in a text much closer to our epoch:

> At the foot of the Garisenda a curious phenomenon was being shown to visitors to the city. If a cloud driven by the wind approached the tower from the leaning side, it seemed that not the cloud, but the tower would move, leaning more and more.[19]

This statement is present in an edition of *Inferno*, published in 1967, by Piero Bargellini. According to what the prolific Florentine journalist and writer – as well as politician – says, the "curious phenomenon" was so frequent and repeatable that it could be shown every day to visitors to the city (perhaps as other mirabilia of the city, like the Neptune fountain in Piazza Maggiore, and the Archiginnasio, as if every day in the sky of Bologna there were fast moving clouds running towards the Garisenda from the slope side). This is undoubtedly false, and it says a lot about the superficiality with which are written works that have wide circulation (such as the numberless editions of the *Commedia* intended for schools or the general public).

From a review of the commentaries on the *Commedia* and other Dante's writings, carried out on printed works and by using electronic resources (e.g., the Dartmouth Dante Project), it emerges that there are very few notes on the two Dante tercets which contain, in relation to the visual phenomenon we are treating, some information useful for the reader. In fact, I managed to find only one, that is present in the edition of *Inferno* by Gabriele Rossetti (the father of the painter Dante Gabriele), published between 1826 and 1827. Rossetti writes commenting on lines 136–145:

> The Carisenda: a tower in Bologna, so called from the name of the person who built it, hangs in such a way that whoever stands below it and looks down at its top, at the point where a cloud passes over it in a direction opposite to the inclination, believes he sees the tower falling above him; this is since, by an optical illusion, it seems to him that the direct motion of the cloud is an opposite motion of the tower. Everyone can experience this in the Moon, which sometimes seems to run towards the clouds, while they are the ones running towards it. Now Dante says that, while he was fully attentive (*a bada*) to see Antaeus bending down to take Virgil, it seemed to him that that the giant was falling on top of him, like Carisenda in the case described.[20]

In this note, which – unlike most editions of the *Commedia* – does not focus on the historical details of the construction, and partial destruction, of the Garisenda, Rossetti correctly places Dante's observation within the class of optical illusions of which the more common is, as we have already said, that of the Moon in apparent movement as the clouds move near or over it.

The short note inserted by Niccolò Tommaseo in his 1837 edition of Dante's masterpiece is almost certainly derived from Rossetti:

> Carisenda: Tower of Bologna, today Torremozza, so leaning, that to those below it would seem as if they saw a cloud passing by it, that not the cloud but the tower was moving. Observation made and retold to me by someone who never read Dante. This is what the Moon does when the clouds move towards it.

Here, despite the extreme brevity of the comment, in addition to the quick reference to the Moon, there is the notation on the fact that an observation similar to that reported by Dante had been made by someone who had never read his poem.

The only other note to the text that I have been able to find, with some useful reflections on the illusory phenomenon, is the one in Giovanni Fallani's 1965 commentary on the *Commedia*:

> When a mobile element (the cloud) approaches another fixed one (the tower) the eye perceives only the relative motion (the approaching of the two points), so that, without other references, it remains difficult to establish which of the elements is in motion and which is stationary.

Here the author, although making no reference to the illusion of the Moon, tries to clarify for the reader how the visual error depends on the absence of other reliable references to which to relate the object of vision.

For the rest, either absolutely nothing is said about the Garisenda illusion, limiting the comments to historical notes on the construction of the tower, or almost only to a paraphrase of Dante's verses, without any real explanatory reflection, or, together with the paraphrase, there are annotations exclusively of a literary or linguistic nature.

Among the modern scholars, Charles Singleton limits his interest to historical notes, particularly on the construction of the tower. Among the earliest commentaries, those by Pietro da Pisa, Ottimo, Graziolo Bambaglioli, Vellutello, and many others are restricted primarily to the paraphrase of Dante's verses. This is also the case of many modern commentators, among whom are noteworthy Pietro Fraticelli in his edition of the *Commedia* of 1902, Allen Mandelbaum in the edition of the *Inferno* published in 1980, Emilio Pasquini and Antonio Quaglio in their edition of the *Commedia* published in 1982, Vittorio Sermonti in the Bruno Mondadori edition of 1991, Robert Durling and Ronald Martinez in their English edition of the *Inferno* published in 1997.

Very short is the notation on the Garisenda present in the Hollanders' edition of the *Inferno* published in 2000:

> The fourth and last simile of the canto refers to one of two towers (the shorter one, in fact, but the one that 'leans' the most) built in Bologna

> in 1109 and 1110. Towers and giants have pride in common, and so the comparison is not without its moral reasons. Its visual reasons are indisputably stunning, a tower that seems to be falling because a cloud is passing over it.

Annotations of a literary or linguistic nature are present in some of the most important modern Italian editions, such as – for example – the one edited by Anna Maria Chiavacci Leonardi between 1991 and 1997, and the one edited by Bianca Garavelli and Maria Corti, which had great success in Italian schools since its first publication in 1993.

Of some interest is the notation of Attilio Momigliano, in his commentary on these Dante's verses in the Sansoni edition of 1957. The Italian literary critic places the emphasis on the phrase "che stava a bada / di vederlo chinare" ("as I watched him bend"), which, in his opinion, underlines the curiosity of Dante, the dominant motif of the whole canto. And – we would add – of the whole *Commedia*, underlining how this curiosity unites Dante the pilgrim of the afterlife world with Dante the author.

Before moving away from this quick review of the comments on the Garisenda in the *Commedia*, we must refer to an interesting element that we find in an edition by Tommaso Casini and Silvio Barbi. It is a long passage taken from a 1909 essay by Ernesto Parodi, in which emphasis is placed on a fundamental theme of this canto, between the "stinking" one of Malebolge and that of the "bleak and frozen land of Cocytus", the reduction to the human sphere of the mythical figures of the classical giants:

> Among the main reasons that inspired Dante is undoubtedly that of reducing the boundless and immoderate imaginations of others, especially of classical poets, within the confines of plausibility, and this motif informs the entire canto. The Poet humanizes the giants. Here his inexhaustible dramatic streak leads to a stupendous intuition: he turns them into beasts as rough and big in mind as in body.[21]

From this perspective, the comparison that Dante establishes between the meek and helpful Antaeus, in the act of bending towards the bottom of the infernal swamp, and the Garisenda which seems to collapse due to the effects of a visual illusion, can be considered an aspect of 'humanization' (and demystification) of the giant (as will be seen in the following verses where his getting back on his feet is compared to the raising of the mast of a ship).

3 With an Eye towards the Sky "When a Cloud Passes"

When, a few years ago, my attention was brought to canto XXI of *Inferno*, my first reaction was to realise that, although I was interested in the study of vision and always looking for illusory effects in the surrounding world – such as Mach bands, blue shadows, the illusions of the Sun and of the Moon – I had never noticed anything immediately similar to Dante's observation. It had certainly happened to me to observe the apparent movement of the Moon in the presence of clouds running across the sky. I also happened to notice something more similar to Dante's illusion. This occurred during a long stay in Minneapolis, a city without leaning towers, but with the three truly fundamental conditions for the observation of this type of phenomena: clear skies, frequently moving clouds, and very tall buildings. It then seemed to me that the tops of the skyscrapers showed a kind of motion, when dense and fast clouds crossed the sky, and this always in the direction of the approaching clouds.

When, after rereading canto XXI of *Inferno*, it had become my habit to turn my eyes to the sky whenever I happened to see moving clouds, an episode happened that really struck me and, in some way, marked the beginning of the intellectual journey that led me to focus my interest on Dante and his visions. I was walking at the foot of a mountain that was not too high, but with steep, almost Dolomite-like walls, when suddenly, an enormous cumulus of clouds appeared from behind the mountain, moving rapidly towards me. I had a rather terrifying impression that the mountain was collapsing, as if there were an earthquake, although I didn't feel any shaking or movement of the earth around me. I must say that the fear aroused by the sudden vision was somewhat attenuated by the fact that, according to the laws governing this type of visual effect, the mountain appeared to collapse on the opposite side to the one we were on.

Since then, for many months I have been trying to study this type of visual illusion. I have made most of the observations during a long stay in the English Midlands, especially in Leicester, which, particularly in the autumn and in winter, offers conditions similar to those of Minneapolis. Here the relative

Figure 4.1 Monte Fammera, the mountain of the Monti Aurunci complex, in the lower part of Lazio region, where I happened to observe the illusory movement in a particularly impressive way. Photograph by the Author.

shortage of skyscrapers was compensated by the proximity, to my residence, of two tall buildings, located right in front of the David Wilson Library of Leicester University. This is the place where I often went to read texts on Dante and his science, and editions of the *Commedia*.

As in Minneapolis, there are no visibly leaning towers or buildings in the English city, but this was not an important handicap, because – as I had discovered quite early in the course of my research – the apparent motion of a terrestrial building, as the clouds move above it, did not require the building to be tilted (unlike what, erroneously, some comments on the *Commedia* suggest). The other thing I immediately realised was that, except in exceptional cases where – due to an extremely rapid movement of the clouds – the effect was particularly impressive, the movement did not usually consist in an apparent collapse, but in a less frightening, horizontal movement of translation.

I did not have the opportunity to verify this by placing myself in the conditions of Dante's observer, that is, by going to Bologna to admire the Garisenda. I would probably have incurred Bassermann's disappointment, unless I had been in the city for a time 'inappropriately' long for me. However, I did it several times in Pisa, with another and no less famous leaning tower, whose building, which was under construction but not completed at Dante's time, was perhaps admired by the poet during one of his stays in the city. The illusion of movement

Figure 4.2 On the left, the two paired tall buildings, located in front of the David Wilson Library of Leicester University and, on the right, the Tower of Pisa, in images taken from videos recorded during the observation of the apparent movement induced by the motion of the clouds. Photograph by the Author.

occurs, substantially in the same way, whether the tower is observed from the side of the slope (the "chinato"), or from the opposite side. It always appears as a translational movement in the horizontal plane, and always in the opposite direction to the motion of the clouds, that is, not like a bending or a collapse.[22]

Depending on whether the clouds move – with respect to the observer – towards the tower, or in the opposite direction, the apparent motion of the building will be either towards or away from the observer.

A surprising thing one discovers when studying this visual illusion is that it is not at all necessary for the building to be particularly tall. It is sufficient that it is high enough, and in such a position that, when observed closely, with the gaze towards its top, other terrestrial objects (situated not underneath the moving clouds) remain out of the vision field. The phenomenon can even be observed for very low buildings, such as the one or two storey Victorian houses quite common in the residential neighbourhoods of Leicester. When it was not possible to exclude the visual background in a natural way, the effect occurred by using some artificial method to narrow the field of vision, for example, by viewing the same scene through a cardboard tube.

The conditions in which the effect tends to appear more impressive depend not only on the speed of the clouds but also on the degree to which they cover the background of the sky. Large piles of white clouds represent the ideal stimulus to evoke the illusion to occur, with one caveat: if the surface of

Figure 4.3 Lincoln Cathedral, at a time when the moving clouds had thinned out and were passing more on the side of the bell tower on the right than on the left (relative to the observer). Photograph by the Author.

the clouds is so uniform that its movement cannot be appreciated, then the phenomenon becomes less impressive or disappears altogether.

By focusing the gaze very carefully on the top of the building, it is possible to observe a very clear effect even when a small cloud passes over the building. This property permits a kind of control experiment that I made during a visit to Lincoln (the city where Grosseteste was bishop), whose splendid cathedral is characterised, like some other Gothic churches, by the presence of two bell towers. In the event that more visible clouds passed over one of the bell towers than over the other, it was the first tower that appeared to be moving more than the second one.

Among the useful, but not indispensable, conditions for the Garisenda illusion to occur is that the terrestrial object to be observed is a building. The phenomenon also occurs if you look at a tree, preferably when it stands alone and extends vertically (like, for example, a cypress). Under these conditions, however, the top of the tree may be shaken by the wind (the same wind that transports the clouds), in which case the effect is more difficult to perceive.

I tried to observe the phenomenon even at twilight. The effect occurs as long as the sky and the building (or tree) observed are clearly visible. It attenuates as the visual contrast decreases with the reduction of ambient lighting. I was able to observe it even on full (and even half-full) moon nights.

There are some interesting variations on this theme that I was able to verify. The first is that of purely celestial observation, of two layers of clouds located at different heights and animated by different movements (or one stationary and the other in movement). Depending on the visual prevalence of the motion of one or the other layer (and also on my effort to concentrate my gaze on one of the two), the layer that was perceptually interpreted as the background appeared stationary, and the other in motion (of course, in the opposite direction).

I was also able to do a similar experiment, albeit with an inverted perspective, from the window of a flying aircraft, when, looking downwards, I saw different layers of clouds moving at different speeds (and far enough away so that their vision was not distorted by the speed of the aircraft).[23]

Finally, it should be noted that, as well as in natural conditions, the illusive movement of buildings or other terrestrial objects also manifests itself – albeit to a lesser extent – when watching video recordings of the observations in the field. For this to happen, even more than in outdoor observations, it is necessary to make an effort to narrow the field of view as much as possible (and to view the recordings on sufficiently large monitors).

The occurrence of the phenomenon in these conditions should not surprise, because the same perceptive mechanisms responsible for the appearance of the illusory motion in natural conditions also operate during the viewing of a video recording (i.e., the relative movement of the object with respect to the background).

4 Concluding Remarks

This essay, in a way an invasion of the field of Dante studies made by a modern scholar of vision, offers the occasion for some reflections.

One concern is the scarcity of attention to the scientific aspects of the Garisenda illusion in the comments on the *Commedia*. As noted, apart from providing a paraphrase of Dante's verses, many texts limit themselves to proposing historical information on the tower and its construction or consist of linguistic and literary analyses. This is certainly understandable, considering the fact most Dante commentators are interested (in addition to making the text comprehensible for readers), mainly in highlighting the literary and stylistic aspects of the poem. This is, of course, praiseworthy because it allows us to penetrate the world of Dante's poetic composition.

Pushed to its limits, in my opinion, this practice risks betraying, in some ways, the spirit of a poem that was written with the ambition of creating a work much more comprehensive than a purely literary poem (at least in the modern sense of the term), a work that included science as one of its important dimensions.

What can Garisenda observation tell us about Dante's scientific culture in the visual field, in particular with regard to his sources?

This is an important problem that has fascinated scholars of varying positions (as is made clear in other chapters of this book). A possibility, supported by some authors, held particularly by Alessandro Parronchi, assumes a poet's in-depth knowledge of optics and vision studies, which in his time were in a phase of important development, especially following the translation, in the first half of the 13th century, of Alhazen's *Kitab al-Manazir*.[24]

A significantly different position is that held particularly by Simon Gilson. The conclusions reached by the English scholar, on the basis of an in-depth and extensive examination of the texts of the time, is that Dante's visual culture was not the expression of direct knowledge of the fundamental texts of medieval *perspectiva* (Alhazen and his Western followers), but that it depended to a large extent on secondary and tertiary sources, which were in fairly wide circulation in the poet's time, and especially on the texts of Albert the Great and Thomas Aquinas – and other works of a philosophical and moral nature.[25]

It is notable however, that, although there are references to illusions or visual deceptions both in the works of Albert and Thomas and in other texts, such as those of Augustine (and in other works), these are so fleeting and little correlated with the specific phenomenon described by Dante.[26] Consequently, they seem not to support the idea that, on the basis of this knowledge alone, the poet could have drawn the elements for his insightful, and probably unprecedented, observation.

Among medieval perspective texts, the most extensive mention of motion illusions is found in Alhazen. Anyone reading the passages on the topic of visual illusions in the third book of his text on optics, and – in particular – his description and analyses of the apparent motion of the Moon in the presence of moving clouds, would have at his disposal the conceptual framework to make an original observation like Dante's on the Garisenda tower.

Among Western perspectives, Bacon is the scholar who, drawing heavily on Alhazen, provides the richest information on the subject (with a clear indication relating to the motion of the moon and stars in the presence of moving clouds), therefore the most effective information to focus the attentive reader's observation skills. Witelo also mentions the Moon in motion in relation to clouds in a very brief passage of his *Perspectiva*, while in the case of Peckham, neither the Moon nor stars are mentioned in the very fleeting notation of his *Perspectiva communis*, in which it is said: "for what concerns the distance, a deception of knowledge and reasoning may happen, whereby moving objects are seen as still, and vice versa".[27]

If we read Dante's Garisenda illusion as a clue to his sources in the field of visual science, then it is difficult to escape the idea that he might have read Alhazen or Bacon, or – at least – that he had a fairly direct knowledge of these texts. He might have acquired this, for example, through contact with scholars interested in the perspective with whom he might have interacted during the time spent in various Italian cities, before and after his exile from Florence.

This, however, is only one possibility. Let's take for granted that visual interest and curiosity was one of Dante's distinctive characteristics, which turned particularly acute in the period in which he was conceiving or creating his masterpiece. Likely he did not limit himself to collecting information from works that he read and heard about, on a private or public level (*lectiones* or homilies). Undoubtedly, he also observed the reality with eyes made more perspicacious by his need to collect, from the real world, elements useful to his aim of humanising a poem which, by its conception and structure, transcended the human.

In addition to the Garisenda illusion, we find other examples of the poet's disposition to see acutely the reality around him and to render it in a rapid and sculptural form. This occurs, for instance, in the simile of the green lizard of *Inferno* XXV, where he visualises metaphorically the speed of the damned (Buoso Donati) who, in the form of an angry little snake, approaches the belly of two companions in misfortune:

> Come 'l ramarro sotto la gran fersa
> dei dì canicular, cangiando sepe,
> folgore par se la via attraversa.
> (*Inf.* XXV, 79–81)

> [As the green lizard beneath the scorching lash / of dog-day heat, between one hedge and the next, / seems lightning as it streaks across the road].[28]

We are tempted to say that, whoever was able to notice and so effectively describe the reptile's rapid motion under the dazzling light of day, very likely had the interest and acuity to grasp, in the world around him, other movements, and – among these – also that of the illusory bending of the Garisenda.

Notes

1 This chapter is the outcome of an event occurred at the time of Covid pandemics, when I happened to listen to a webinar on Dante given by a friend of my times of studentship in medicine in Rome, Nicola Baldi. In his talk, Nicola impressed me by revealing how much medicine is present in the *Commedia* and in other works by Dante. In particular, I was surprised to hear that the Italian poet also described a visual illusion. This was the beginning of an exciting period of studies and discussions with colleagues and scholars that helped me to revive the interest in Dante, dormient since my schooldays. This is why I wish to dedicate this chapter to Nicola.

2 For this issue, see, in this volume, Piccolino, "Vision as a Tangible and Dynamic Tool". The quotations of Dante' *Commedia* are from the edition by Petrocchi and from the English edition by Robert and Jean Hollander, both available at the Danteonline site: https://www.danteonline.it/ (where, however, the title of the translation is incorrectly spelt as *Divine Commedy*). From the same site, I have also taken the quotations from the *Convivio*, in the Italian edition of Brambilla Ageno, and in the English translation by Lansing.

3 This theme is also dealt with by Galli, "Visual Perception", in this volume.

4 The frequent use in the *Commedia* of scenes of uncertain or deceptive vision, due to the distance, particularly in dimly lit environments (as in *Inf.* XV, 20–21, where the souls of the lustful "sharpened their eyelashes / as the old tailor does in his eye"), has raised the possibility of an "ophthalmic defect" of the poet (Wlassics, *Dante narratore*). Whatever the case, we need to think about how uncertain vision dominated the collective imaginary in a period in which the use of glasses to correct visual anomalies was extremely limited. With relation to the deceptive appearance of towers seen from afar, particularly expressive is Lucretius' *De rerum natura*, IV, 353–356, where the poet says that this occurs because each distant angle is perceived obtuse, or rather it is not perceived at all ("Quadratasque procul turris cum cernimus urbis, / propterea fit uti videantur saepe rutundae, / angulus optusus quia longe cernitur omnis / sive etiam potius non cernitur").

5 "Dove è da sapere che, propriamente, è visibile lo colore e la luce, sì come Aristotile vuole nel secondo de l'Anima, e nel libro del Senso e Sensato. Ben è altra cosa visibile, ma non propriamente, però che altro senso sente quello, sì che non si può dire che sia propriamente visibile, né propriamente tangibile; sì come è la figura, la grandezza, lo numero, lo movimento e lo stare fermo, che sensibili [comuni] si chiamano: le quali cose con più sensi comprendiamo. Ma lo colore e la luce sono propriamente [sensibili]; perché solo col viso comprendiamo ciò, e non con altro senso". This passage from the *Convivio* and in particular the mention of light, together with colour, as proper sensible (not present in Aristotle but present in Alhazen and in the perspectivist literature), has led Cecilia Panti – together with other considerations – to suggest an orthographic correction (See Panti, "Il visibile per retta linea" and Panti, "Seeing the Light", in this volume). In her reconstruction, the phrase reads: "Here we should know that properly speaking only color and light are visible. As Aristotle asserts in the second book of *On the Soul* and in *On Sense and Sensibles*, it is true that other things are visible, but not properly speaking [...]".

6 "E sì come peregrino che va per una via per la quale mai non fue, che ogni casa che da lungi vede crede che sia l'albergo, e non trovando ciò essere, dirizza la credenza a l'altra, e così di casa in casa, tanto che a l'albergo viene; così l'anima nostra, incontanente che nel nuovo e mai non fatto cammino di questa vita entra, dirizza li occhi al termine del suo sommo bene, e però, qualunque cosa vede che paia in sé avere alcuno bene, crede che sia esso".

7 We note here, *en passant*, the use of the cinematographic "point of view" shot dealt with in Piccolino, "Vision as a Tangible and Dynamic Tool" in this volume.

8 Gilson, *Medieval Optics*, 100.

9 Notice Dante's spelling of the tower name, which, might possibly have been more current than Garisenda in the Tuscan language of his time.

10 The beauty and conciseness of these verses have been remarked in a particularly expressive way by Antonio Cesari in his commentary on Dante's *Inferno* written in the form of a dialog and published in 1824 (Cesari, *Bellezze della Commedia*, 217). In his youthful years, Dante might have dedicated a fairly famous sonnet to the Garisenda, which also refers to a visual deception, although of a very different nature. Although still a matter of debate, it is indeed probable that Dante stayed in Bologna. This possibly occurred the first time, between 1293 and 1294, when the poet was in his twenties, and, afterwards, at the time of his exile, starting from 1304. See Tavoni's edition of *De vulgari eloquentia*, and Tavoni, "Quanto è probabile".

11 See Wade, *A Natural History of Vision*, and Wade, *Perception and Illusion.* See also the chapter by Wade, "Visual Motion Illusions", in the present volume.

12 A particularly impressive variation of this experiment can occur to an observer situated inside the cabin of an airplane when he looks outside through one of the windows of the aircraft. If, by chance, a moving object happens to occupy the scene seen through the window (for instance, a bus for passengers' transport or an embarkation ladder), then the observer feels as if the aircraft is moving in the opposite direction. The particularity of this impression is in the fact that a huge vehicle can appear to be moving when the perceived motion occupies a relatively restricted view, as that of the small cabin window.

13 Alhazen, *Optics*, III, 260–261.

14 In a recent essay on the "reference system" in Dante, Ottavio Brigandì has highlighted an instance of illusory motion in Lucan's *Pharsalia.* In this instance, both the real moving background and the visual object, whose illusive motion is perceived, are terrestrial objects: the sudden motion of the sea in the foreground makes sailors believe that the distant mountains are moving (Brigandì, "Il cambio di sistema di riferimento").

15 De Brosses, "Lettres familières", 232-248 (XX. A M. de Neuilly. Memoire sur Bologne): 235.

16 Goethe, *Viaggio in Italia*, 104. It is possible that at the time of these notes by Goethe on the Garisenda (18 October 1787), the German poet had not read the *Commedia.* This could explain his lack of interest in the illusory motion of the leaning tower as the clouds pass. Note, on the other hand, that at the time of the trip to Italy Goethe was very interested in science, and there are many places of medical-scientific interest in this country that he visited with interest. Among other things, during his stay in Venice in 1784 he discovered the intermaxillary bone on the beach of the Lido which led him to formulate some theories of comparative morphogenesis.

17 Goethe, *Viaggio in Italia*, 109. A mention of the leaning towers of Bologna is present in the accounts of many personages who have visited Italy in the course of the centuries, but in none of those accounts, I have been able to find an observation comparable to that described by Dante for the Garisenda. Among those

accounts are notable those of Michel de Montaigne, who was in Bologna in 1580, of Maximilien Misson, who visited the town in 1688, of Madame de Staël who wrote about the Garisenda in 1816, of John Ruskin who visited Bologna in two occasions in the first half of the 18th century.

18 Bassermann, *Orme di Dante in Italia*, 213.

19 Bargellini, "Commento a *Inferno*", 385.

20 Rossetti, "Commento a *Inferno*", as most of the early commentaries to the *Commedia*, have been consulted at the site of the Darmouth University: https://dante.dartmouth.edu. This and the two subsequent translations are mine.

21 Parodi's note in *Divina Commedia*, edited by Casini and Barbi, 801.

22 Also in Pisa I had recently the opportunity to make one observation with another leaning tower, the Campanile of the Church of San Michele degli Scalzi, situated near the Arno River at a short distance outside the medieval walls of the town. This tower, which, as the Garisenda has a squared section, is about half the height of the Bologna tower but has a more pronounced tilting (5° versus 4° of the Garisenda). The illusive motion produced by the moving clouds was quite impressive.

23 Another personal observation, perhaps of a not very relevant nature in the context of a study on Dante, is that, when the phenomenon is observed from the ground for a long time, at a certain point something happens that makes us risk falling, and in a real and non-illusory way, in the opposite direction to the movement of the clouds It is likely that this is due to a perceptual adaptation that makes the illusory visual input prevail over proprioceptive sensations, leading to an altered perception of our position in space (a mutation of the spatial reference system induced by the anomaly in the visual sensation).

24 See Parronchi, *Studi su la dolce prospettiva*.

25 Gilson, *Medieval Optics*. It is singular, however, that this extremely comprehensive text does not take the Garisenda illusion into consideration. Among modern authors, it seems that only Brigandì, "Il cambio di sistema di riferimento", highlights the importance of the observation of the Garisenda in the conceptual world of the poet. The phenomenon is mentioned in two interesting articles, respectively by Marcello Ciccuto and Massimo Colella, but in no case there is a reflection on the underlying perceptual mechanisms. In Ciccuto, *Scoto e la* naturalis philosophia *di Dante*, 34, this happens even though the author insists on the "deception of perception" and the "frustration of senses" implicit in the illusory fall of the tower, as in the apparent fall of Anthaeus. On his hand, Colella, "Simili modo poete ponunt gygantes", 95, deals with the ancient commentators of the *Commedia*, and appropriately highlights, in the parallel established, in the canto, between giants and towers, the fact that the Garisenda stays to Anthaeus, as the Babel tower stays to Nimrod and a high tower shaken by an earthquake stays to Ephialtes.

26 From a recension of the texts of these authors, to which Dante might have some access, the most significant passage I have been able to find is in Albert the Great, *De Sensu et Sensato*, I, 1, XIV, 51–52: "Adhuc autem diu motus navi vel alio vehiculo si convertat visum ad terram, videtur ei terra moveri cujus equidem eadem causa est, quia servatur in visu et spiritu visivo forma motus". Here the author invokes an example of induced motion, present in different forms in Aristotle, Lucretius, and Ptolemy, as an example of the persistence of visual sensation after the cessation of the stimulus. Another short passage, signalled to me by Simon Gilson, is in Augustine, *De genesi ad litteram*, XII, 25, where the author states that sensible vision can be deceptive in leading to the perception of movement, when the visual objects are moving, and viceversa: "fallitur ergo in visione corporali, cum in ipsis corporibus fieri putat, quod fit in corporibus sensibus -

sicut navigantibus videntur in terra moveri quae stant, et intuentibus caelum stare sidera quae moventur". In no simple way, these passages seem to provide the intellectual framework for the intuition of an illusion similar to the Garisenda'.

27 Peckham makes no mention of the phenomenon in his earlier and more extensive *Tractatus de perspectiva.*

28 The scientific explanation of the extraordinary speed of this reptile, rather noteworthy for its large size, compared to the more common lizards, and for its showy green colour, is related to its nature as a formidable daytime predator. This characteristic implies a great speed of movement in the phase of capturing prey and in moving between the areas in which it hides itself to prepare its ambushes, particularly during the full light of sunny days (and also due to the need not to be spotted by predators, as well as by prey).

Bibliography

Primary Sources

Alhazen. *Optics.* Translated and edited by Abdelhamid Ibrahim Sabra. *The Optics of Ibn Al-Haytham, Books 1–3.* London: Warburg Institute, 1989.

Albert the Great. *De sensu et sensato.* Edited by Silvia Donati. Münster: Aschendorff, 2017.

Augustine. *De genesi ad litteram libri duodecim.* Edited by Jacques Paul Migne. *Patrologia Latina,* vol. 34, coll. 245–486. Paris: J. P. Migne, 1865.

Dante Alighieri. *Convivio.* English translation by Richard H. Lansing. *Dante's "Il Convivio".* New York: Garland, 1991.

Dante Alighieri. *Convivio.* Edited by Franca Brambilla Ageno. Florence: Le Lettere, 1995.

Dante Alighieri. *La Divina Commedia.* Edited by Tommaso Casini and Silvio Adrasto Barbi [6th edition]. Florence: Sansoni, 1959.

Dante Alighieri. *The Divine Comedy, Inferno, Purgatorio, Paradiso.* Translated by Robert Hollander and Jean Hollander. New York: Doubleday, 2003, 2007.

Dante Alighieri. *De vulgari eloquentia.* Edited by Mirko Tavoni. Milan: Mondadori, 2017.

Dante Alighieri. *La Divina Commedia.* Edited by Giorgio Petrocchi. Florence: Le Lettere, 1994.

Secondary Sources

Bargellini, Piero. "Commento" to Dante Alighieri. *La Divina Commedia. Inferno.* Florence: Vallecchi, 1967.

Bassermann, Alfred. *Orme di Dante in Italia: vagabondaggi e ricognizioni.* Translated and edited by Egidio Gorra. Bologna: Zanichelli, 1902.

Brigandì, Ottavio. "Il cambio di sistema di riferimento quale aspetto innovativo della riflessione gnoseologica di Dante". In *Letteratura e Scienze*, edited by Alberto Casadei, Francesca Fedi, Annalisa Nacinovich, and Andrea Torre, 1–4. Rome: Associazione degli Italianisti, 2021.

Cesari, Antonio. *Bellezze della Commedia di Dante Alighieri. Inferno.* Verona: Libanti, 1824.

Ciccuto, Marcello. "Michele Scoto e la *naturalis philosophia* di Dante". *Tenzone* 4 (2003): 31–41.

Colella, Massimo. "'Simili modo poete ponunt gygantes'. Sull'antica esegesi di *Inf.* XXXI". *Rivista di studi italiani* 39/3 (2021): 80–127.

De Brosses, Charles. *Lettres familières écrites d'Italie en 1739 et 1740*, edited by Roman Colomb, vol. 1. Paris: Librairie Académique P. Didier et Co., 1858.

Fallani, Giovanni. "Commento" to Dante Alighieri. *La Divina Commedia. Inferno.* Edited by Giovanni Fallani. Messina and Florence: G. d'Anna, 1965.

Gilson, Simon A. *Medieval Optics and Theories of Light in the Works of Dante.* Lewiston, Queenston and Lampeter: The Edwin Mellen Press, 2000.

Goethe, Johann Wolfgang. *Viaggio in Italia: 1786–1788.* Edited by Eugenio Zaniboni. Florence: Sansoni, 1959.

Panti, Cecilia. "Il visibile 'per retta linea'. Dante, la perspettiva e una proposta di emendamento in *Convivio* III, ix, 6". In *Dante e il mondo: tra realtà e poesia tra storia e letteratura. Atti del LVIII Convegno storico internazionale. Todi 10-12 ottobre 2021*, 143–180. Spoleto: Centro Italiano di Studi sull'Alto Medioevo, 2022.

Parronchi, Alessandro. *Studi su la dolce prospettiva.* Milan: Martello, 1968.

Rossetti, Gabriele. "Commento" to Dante Alighieri. *La Divina Commedia*, edited by Gabriele Rossetti, vol. 1. London: John Murray, 1826.

Tavoni, Mirko. "Quanto è probabile che Dante abbia scritto il *De vulgari eloquentia* a Bologna e perché ci interessa?". *Nuova Rivista di Letteratura Italiana* 24/2 (2021): 11–109.

Tommaseo, Niccolò. "Commento" to Dante Allighieri. *La Commedia. Col comento di N. Tommaseo*, vol. 1. Venice: Il Gondoliere, 1837.

Wade, Nicholas James. *A Natural History of Vision.* Cambridge, MA: MIT Press, 1998.

Wade, Nicholas James. *Perception and Illusion. Historical Perspectives.* New York: Springer, 2005.

Wlassics, Tibor. *Dante Narratore. Saggi sullo stile della* Commedia. Florence: Leo S. Olschki, 1975.

5 Where Do Visions That Do Not Come from Sight Come from?

Mirko Tavoni

1 Introduction

Dante is animated by a powerful visionary inclination throughout his life and draws inspiration from it at various stages of his history as a poet. This inclination manifests itself in the most evident way in his youthful work, the *Vita Nova*, all crossed by dreams and visions, and in the masterpiece of his maturity, the *Divine Comedy*, which is obviously a vision of the afterlife, and not a fictitious one, rather a vision that Dante swears several times to have truly had, and of which several episodes reveal a semi-hidden oneirism of impressive originality.

Dante wrote the *Vita Nova* in the years around 1292–1295, and the *Divine Comedy* between around 1307 and his death in 1321. The intermediate period between the one and the other is the only one of his life in which Dante does not dream, has no visions, and does not draw on this psychic dimension for his literary writing. In the first, terrible years of exile, starting from 1302, he is too busy fighting, trying to survive, and defending his side politically and militarily to be able to allow himself to dream and fantasise. And when he regains the possibility of living in relative peace, of reading books and thinking, first in Verona then in Bologna, from 1303 to 1304 onwards, he devotes all his cultural commitment to writing the *Convivio*, an imposing treatise on Aristotelian philosophy in the vernacular at the service of the illiterate nobles who governed many cities in northern Italy.

In a passage of this work (II, viii, 7–13) Dante explains, in philosophical and theological terms, where divinatory dreams come from. It is a key passage in the evolution of his thought, because here he shows that he found in his philosophical-theological sources – Albert the Great but also Avicenna – the explanation of all the phenomena of 'inner vision' that were so important in the experience of love for Beatrice which he had narrated in poetry and prose in the *Vita Nova*. And these authoritative sources reveal to him that these psychic phenomena are not pathological, are not signs of a weak personality, but on the contrary are privileged phenomena, reserved for strong personalities, of great nobility of soul, and endowed with prophetic value.

DOI: 10.4324/9781003412601-6

Dante, thus, becomes aware that his own special visionary inclination has a high cognitive value. This is what gives him a strong impulse to launch himself into the daring and never-before-attempted undertaking of reporting a supreme, entire vision of the afterlife. In fact, Dante will repeat the same philosophical-theological explanation exposed in the *Convivio* in canto XVII of *Purgatorio* (vv. 13–18), as a comment on the ecstatic visions he is having while walking in trance along the third and fourth cornice (ring) of the mountain of Purgatory.[1]

2 Dreams and Visions in the *Vita Nova*

The dreams and visions of the *Vita Nova* are not eight, as they apparently are, but nine, because at the centre of the narrative, there is an encrypted vision, the one listed here at n. 7.

These dreams and visions are listed below.[2]

2.1 Vita Nova *1, 2–11 (II, 1–10)*

The famous, fatal first meeting with Beatrice at the age of nine is narrated here (1, 2–4). The meeting takes place in reality, but we note the insistence of the verb *apparire*:

> to my eyes first appeared the glorious lady of my mind [...] at about the beginning of her ninth year did she appear to me [...] She appeared humbly and properly dressed in a most noble color, crimson.
>
> [1, 2–4: a li miei occhi apparve prima la gloriosa donna de la mia mente [...] quasi dal principio del suo anno nono apparve a me [...] Aparve vestita di nobilissimo colore, umile ed onesto sanguigno].

Apparire and *parere* are the epiphanic verbs that dominate both in real scenes and in scenes resulting from *imaginazione*. The woman, as in Cavalcanti, is an apparition already when she enters the scene in reality, then she is an apparition, again, when she returns in fantasy. There is continuity between what is seen with the eyes and what is seen only in the mind. External or internal, it is the same type of peremptory view that imposes itself.

Beatrice's appearance then triggers (1, 5–7) a typical stilnovo psychodrama: the spirit of life, which *trembles*, the animal spirit, which *is amazed*, and the natural spirit, which *cries*, each utter a line in an oracular Latin, respectively: *Ecce deus fortior me, qui veniens dominabitur michi* ("Behold a god more powerful than I, who comes to rule over me"); then, addressed to the spirits of sight, *Apparuit iam beatitudo vestra* ("Now has appeared your beatitude"); third, *Heu miser, quia frequenter impeditus ero deinceps!* ("Wretched me, for often hereafter shall I be impeded!").

This psychodrama takes place in Dante's mind when he is awake, the moment he sees Beatrice for the first time, but it has the same characteristics as the psychodrama that will take place in his dream nine years later, here in Section 2.2. And from that moment Love (the god stronger than the spirit of life), who here appears personified for the first time, took control of Dante's soul and

> he began to assume over me such assurance and such mastery, through the power that my imagination gave him, that I was obliged to do all his bidding fully.
>
> [1, 8: cominciò a prendere sopra me tanta sicurtade e tanta signoria, *per la vertù che li dava la mia imaginazione*, che mi convenia fare tutti li suoi piaceri compiutamente].

2.2 Vita Nova *1, 12–18 (III, 1–7)*

Here we have the second fatal *apparimento* of Beatrice – in reality, on the street – nine years after the first:

> this marvelous lady appeared to me dressed in purest white, between two gentle ladies who were of greater years; and passing along a street, she turned her eyes to that place where I stood in great fear.
>
> [1, 12: questa mirabile donna *apparve* a me vestita di colore bianchissimo, in mezzo a due gentili donne le quali erano di più lunga età; e, passando per una via, volse gli occhi verso quella parte ov'io era *molto pauroso*].

Beatrice's "dolcissimo salutare" inebriates Dante to the point that he has to retreat:

> to the solitary place of my room, and set about thinking upon this most courteous one. And thinking of her, I was overcome by a sweet sleep, in which there appeared to me a marvelous vision.
>
> [1, 13–14: al solingo luogo d'una mia camera, e puosimi a pensare di questa cortesissima; e pensando di lei mi sopragiunse un soave sonno nel quale m'apparve una maravigliosa visione].

It should be noted that the dream is preceded and prepared by intense *pensare*: the osmosis between thought and dream will remain a constant in Dante; just think of the line of *Purgatorio* XVIII, 135 "e 'l pensamento in sogno trasmutai".

The protagonist of the dream vision is "of an aspect frightening to whoever might behold him" ("una figura d'un segnore di pauroso aspetto") who speaks in Latin like the spirits of the psychodrama experienced while awake referred to in Section 2.1 above, and is confirmed to be *Amore*, the god who nine years earlier had become lord of Dante's soul, and which here reiterates "I am your Lord" (1, 14: *Ego dominus tuus*).

In a development of "it seemed to me" ("parea", "pareami"), a dreamlike scene unfolds, mixing together fear (*paura*) and sweetness (*dolcezza*). Just as in the meeting with Beatrice, Dante is "very fearful" ("molto pauroso") and Beatrice's *salutare* is *dolcissimo*, so in the dream the figure of Love is "of an aspect frightening to whoever might behold him" ("di pauroso aspetto a chi la guardasse"), but at the same time, "he appeared so happy in himself that it was a wondrous thing" (1, 14: "pareami con tanta letizia, quanto a sé, che mirabil cosa era"). The scene is highly erotic:

> I seemed to see in his arms a person asleep, naked except that she seemed to me lightly wrapped in a crimson cloth; one whom I, regarding with great attention, recognized as the lady of the salutation, who the day before had deigned to greet me.
>
> [1, 15: Nelle sue braccia mi parea vedere una persona dormire nuda, salvo che 'nvolta mi parea in un drappo sanguigno leggeramente; la qual i' guardando molto intentivamente, conobbi ch'era la donna de la salute, la quale m'avea lo giorno dinanzi degnato di salutare].

It develops into an anguished carnal outcome, as Love wakes Beatrice and forces her to eat Dante's heart which she holds in her hand: *Vide cor tuum* (1, 16: "Behold your heart"). Then, the joy of Love:

> converted into most bitter weeping; and thus weeping, he gathered this lady into his arms, and with her he appeared to ascend toward heaven, for which I suffered such great anguish that I could not sustain my fragile sleep, which indeed broke and I awakened.
>
> [1, 18: si convertia in amarissimo pianto e, così piangendo, si ricogliea questa donna nelle sue braccia e con essa mi parea che si ne gisse verso il cielo; ond'io sostenea sì grande angoscia che 'l mio deboletto sonno non poteo sostenere, anzi si ruppe e fui isvegliato].

The ascent to heaven seems to prophesy Beatrice's death.

2.3 Vita Nova *4, 1–8 (IX, 1–8)*

Dante is riding, together with a company of other knights, away from Florence, and is distressed because he is moving away from Beatrice. While he's riding, Love

> appeared in my imagination as a pilgrim dressed meagrely and in simple vestments. He seemed to me frightened, and he stared at the ground, except that at times his eyes seemed to turn toward a river, beautiful and swift and utterly clear, that flowed along the road there where I was.

> [4, 4: ne la mia imaginazione apparve come peregrino leggeramente vestito e di vili drappi. Elli mi parea disbigottito e guardava la terra, salvo che talora li suoi occhi mi parea che si volgessero ad un fiume bello e corrente e chiarissimo lo quale sen gìa lungo questo cammino là ov'io era].

Dante, fully awake, experiences a sort of "augmented reality": he sees everything that surrounds him, the moving scene of his ride, but furthermore, within it, produced by his *imaginazione*, he sees Love and interacts with him. Love, in fact, speaks to him, gives him an articulate and lucid speech, full of details, and then suddenly disappears.

Imaginazione is here both the faculty in which the vision takes place (4, 3: "ne la mia imaginazione apparve come peregrino") and the vision itself: "this imagining of mine disappeared" (4, 7: "disparve questa mia imaginazione"). In the *Vita Nova* only the word *imaginazione*, the word of the common lexicon, is present (and it is very frequent), while the word *imaginativa*, a philosophical technicality, is not, because Dante has not yet discovered the philosophical concepts relating to imagination. As for Love, it is already the third time that he appears: three out of three in the first visions.

2.4 Vita Nova *5, 8–16 (XII, 1–9)*

Beatrice denies Dante "her very sweet greeting in which all *his* bliss was" (5, 2); so he, gone "in a lonely place [...] to wet the earth with the bitterest tears" (8), withdraws in his room, "where *he* could complain without being heard"; and there, calling for mercy to the lady of courtesy and saying "Love, help your faithful one", fell asleep "like a beaten little child crying" (9). And here, a *visione* (so called by Dante) occurs again in a dream. Again, Love appears to him, this time in the form of "a young man dressed in very white robes" who sits on his bed, and again (among many *pareami*, *parvemi* and the like) speaks to him in Latin (5, 10): *Fili mi, tempus est ut pretermictantur simulacra nostra* ("My son, it is time to end our fabrications"); and even more obscurely (5, 11): *Ego tamquam centrum circuli cui simili modo se habent circumferentie partes, tu autem non sic* ("I am like the center of a circle, to which all the points of the circumference bear the same relation; you, however, are not").

I will not discuss here which may be the 'sources' of the words spoken by Love and other subjects in these visions, as well as of the appearance, the clothing, and the gestures of the various characters who act in the visions. Several scholars have practiced this type of analysis and have produced intertextual comparisons between the details of Dante's text and those of other texts, without asking and implicitly considering irrelevant, whether the possible reuse by Dante of such contents was or was not mediated by a dream-visionary experience, which Dante firmly declares to have existed. To me, however, the existence of this dream-visionary experience appears self-evident, and an essential fact to recognise, in order to understand what kind of text the *Vita Nova* is.

Love, after uttering his obscure oracular sentences in Latin, descends to speak in the vernacular, and explains to Dante why Beatrice took away his greeting and gives him practical instructions on how he must write a certain message for her in rhyme.

2.5 Vita Nova *14, 1–16 (XXIII, 1–16)*

This long passage tells us about an *imaginazione*, or *fantasia*, induced by "a dolorous illness" (*dolorosa infermitade*). The suffering and debilitation brought by this *dolorosa infermitade* are very strong:

> from which continuously for nine days I suffered bitterest pain, which reduced me to such weakness that I was obliged to lie down like those who cannot move [...] feeling myself in pain almost intolerably [...] returned to thinking about my enfeebled life, and to seeing how fleeting was its duration, even when it was healthy, I began to weep within myself over such misery.
>
> [14, 1–3: onde io continuamente soffersi per nove dì amarissima pena, la quale mi condusse a tanta debolezza che mi convenia stare come coloro li quali non si possono muovere; [...] sentendome dolere quasi intollerabilemente; [...] io ritornai pensando a la mia deboletta vita; e veggendo come leggiero era lo suo durare ancora che sana fosse, cominciai a piangere fra me stesso di tanta miseria].

Against this background the thought of death comes forward: of the death of Dante himself, and of that of Beatrice: "Hence, sighing heavily, I said within myself: 'Of necessity it must happen that the most gentle Beatrice one day will die'" (14, 3: "Onde, sospirando forte, dicea fra me medesimo: 'Di necessità conviene che la gentilissima Beatrice alcuna volta si moia'").

Physical prostration triggers a frightening hallucination, in which the *imaginatio*, detached from *intellectus* and *cognitio* as a result of the infirmity weakening the body, produces sensorial contents that appear to the subject as externally objectified. This is exactly what happens to Dante in this "erronea fantasia" (14, 8), which may seem to happen in his sleep, and therefore to be a dream, but is more than a dream; it is a terrible *farneticazione*: "came over me so powerful a sense of loss that I closed my eyes and began to be agitated like a frantic person" (14, 4: "mi giunse un sì forte smarrimento che chiusi gli occhi e cominciai a travagliare sì come farnetica persona").

This hallucination, or ranting, or delirium, whatever we want to call it, makes the body of the dead Beatrice appear in Dante's eyes (14, 8) and culminates in a macabre exploration – after which it spills into a celestial vision:

> And when I had seen completed all the doleful ministrations that the bodies of the dead it is customary to perform, I seemed to return to my

> room and there to look toward heaven; and so strong was my imagination that, weeping, I began to utter in my real voice: "Oh, most beautiful of souls, how blessed is the one who sees you!".
>
> [14, 10: E quando io avea veduto compiere tutti li dolorosi mestieri che a le corpora de' morti s'usano di fare, mi parea tornare nella mia camera e quivi mi parea guardare verso lo cielo; e sì forte era la mia imaginazione che, piangendo, incominciai a dire con verace boce: "Oi, anima bellissima, come è beato colui che ti vede!"].

It is important to note that Dante repeatedly qualifies this *fantasy* as *erroneous* (the Italian verb *errare* ranges from the meaning of 'wander' to that of 'make a mistake', 'be wrong'):

> at the outset of the wandering that my fantasy took to [...] Thus, my fantasy beginning to wander [...] and so strong was the erroneous fantasy that it showed me that lady dead [...] and I had already said "Oh, Beatrice" when, retuming to myself, I opened my eyes and saw that I was deceived [...] Hence I, being somewhat comforted and aware of my false imagining [...] After this empty imagining [...].
>
> [14, 4–15, 1: nel cominciamento de *l'errare che fece la mia fantasia*; (5) Così cominciando ad *errare la mia fantasia*; (8) e fue sì forte *la erronea fantasia* che mi mostrò questa donna morta; (13) E già detto avea "O Beatrice", quando, riscotendomi, *apersi li occhi, e vidi ch'io era ingannato*; (15, 1) onde io, essendo alquanto riconfortato, e conosciuto *lo fallace imaginare*; Appresso *questa vana imaginazione*].

This is the most precise – intolerably precise – divinatory vision, or prophecy if we want to call it, of Beatrice's death, the central event of the *Vita Nova*, on which the whole story gravitates; but despite being tremendously truthful, Dante insists on qualifying it as *erroneous*, *fallacious*, *deceptive*. We will have to comment on this contradiction.

2.6 Vita Nova *15, 1–6 (XXIV, 1–6)*

Love appears again, to an awake Dante, in the streets of Florence, as in the ride here in Section 2.2. Here too we could define this experience as 'augmented reality', in which the character Love, fruit of the imagination, acts together with the real characters Beatrice and Giovanna "Primavera", the woman loved by Dante's "first friend", the poet Guido Cavalcanti.

This is the episode in which Beatrice's assimilation to Jesus is most explicit, alongside Giovanna, who turns out to have this name because she plays the role of John the Baptist who prepares the advent of Beatrice-Jesus. This meaning is also expressed by Giovanna's nickname, *Primavera*, so called because she "will come first" (*prima-verrà*), alluding to the fact that Cavalcanti the poet prepares the way for Dante the poet. This whole constellation of meanings is interpreted and declared by the character Love.

2.7 Vita Nova *19, 1–2 (XXVIII, 1–2)*

This is the long-awaited and long-prepared moment of Beatrice's death:

> the Lord of Justice called this most gentle one to glory under the sign of the blessed queen, the Virgin Mary, whose name was in greatest reverence in the words of this blessed Beatrice.
>
> [19, 1: lo Signore de la giustizia chiamòe questa gentilissima a gloriare sotto la 'nsegna di quella reina benedetta virgo Maria, lo cui nome fue in grandissima reverenzia nelle parole di questa Beatrice beata].

And here, surprisingly, Dante tells his readers that he does *not* intend to deal with Beatrice's death, that is, the central event of the whole story:

> And although it might perhaps be pleasing at this point to treat somewhat her departure from us, it is not my purpose to deal with it here for three reasons: the first is that it is not part of the present purpose, if we want to consider the proem that precedes this little hook; the second is that, supposing it did pertain to the present purpose, my language would still be inadequate to deal with it properly; the third is that, given the first and second conditions, it is not becoming for me to treat of it, for the reason that, by treating it, I would be obliged to be a praiser of myself, which thing is at all times reprehensible to whoever does it; and therefore I leave this matter to some other glossator.
>
> [19, 2: E, avegna che forse piacerebbe a presente trattare alquanto de la sua partita da noi, non è lo mio intendimento di trattarne qui per tre ragioni. La prima è che ciò non è del presente proposito, se volemo guardare nel proemio che precede questo libello; la seconda si è che, posto che fosse del presente proposito, ancora non sarebbe sofficiente la mia lingua a trattare come si converrebbe di ciò; la terza si è che, posto che fosse l'uno e l'altro, non è convenevole a me trattare di ciò, per quello che, trattando, converrebbe essere me laudatore di me medesimo, la qual cosa è al postutto biasimevole a chi lo fae, e però lascio cotale trattato ad altro chiosatore].

Here, apparently, there is no vision. Instead, there is a vision, encrypted. Dante is telling us, in such a cryptic way that he was not understood for seven centuries, that, at the exact moment of Beatrice's death, he followed her soul to heaven up to God. That is, he tells us that he lived the same experience of *raptus* to the third heaven that Saint Paul recounts in the *Second Letter to the Corinthians*. The three mysterious reasons he gives for *not* telling Beatrice's death mean: (1) that this episode, being an episode of mystical abduction, exceeds the limits of his mind and is therefore not recorded in the "book of memory", the content of which is the purpose of *Vita Nova* to copy; (2) that his language is not ready to deal with such an ineffable experience; (3) that, if he were to report this episode, he would run the risk of "being self-praising", just as Saint Paul says that he delayed reporting his seizure for fourteen years so that it would not appear that he wanted to "glory" about it. This is therefore the true, first occurrence, in real time, of the final *mirabile visione*. Dante is already aware that he is emulating St. Paul's *raptus* to the third heaven. This is the turning point, compared to the previous visions, which were all in Beatrice's life, all born of the love relationship (even if profoundly distorted from its distant courtly origin) and in fact all dominated by the character of Love. Thanks to this encrypted vision, the "inner visions" of the *Vita Nova* – whether in sleep or while awake – are not eight, an insignificant number, but nine, the true key number of the *Vita Nova*, the number that identifies Beatrice:

> this lady was accompanied by this number nine to make it understood that she was a nine, that is, a miracle, whose root – that is, of the miracle – is solely the wondrous Trinity.
>
> [19, 6 (XXIX 3): questa donna fue accompagnata da questo numero del nove a dare ad intendere ch' ella era uno nove cioè uno miracolo, la cui radice, cioè del miracolo, è solamente la mirabile Trinitade].

2.8 Vita Nova *28, 1 (XXXIX, 1)*

Here we are in the midst of the conflict between Beatrice's memory and the temptation of the new love for the "gentle woman", the woman who shows mercy to Dante and with whom Dante risks falling in love:

> Against this adversary of reason [that is, the temptation of a new love for the "gentle woman"] arose in me one day, at about the ninth hour, a powerful imagining: that is, I seemed to see this glorious Beatrice with those crimson vestments in which she first appeared before my eyes; and to me she seemed young and of the same age as when I first saw her.
>
> [Contra questo aversario de la ragione si levòe un die, quasi nell'ora della nona, una forte imaginazione in me, che mi parve vedere questa gloriosa Beatrice con quelle vestimenta sanguigne colle quali apparve prima a li occhi miei; e pareami giovane in simile etade ne la quale io prima la vidi].

This daytime, open-eyed *imaginazione*, which brings Beatrice back on stage with the blood-red dress she wore in her first apparition at the age of nine (Section 2.1) and in her second, erotic dream-apparition at the age of 18

(Section 2.2), and the thoughts that follow from this *imaginazione*, abstract Dante from the surrounding world to the point that he forgets where he is.

2.9 *And Finally,* Vita Nova *31, 1–3 (XLI, 1–3)*

> After this sonnet [*Oltre la spera che più larga gira*] there appeared to me a wonderful vision, in which I saw things that made me resolve to write no more of this blessed one until I could more worthily treat of her. And to arrive at that, I apply myself as much as I can, as she truly knows. So that, if it be pleasing to Him for whom all things live that my life may last for some years, I hope to say of her what was never said of any other woman. And then may it please Him who is the Lord of courtesy that my soul may go to see the glory of his lady: namely, that blessed Beatrice, who in glory gazes upon the face of Him *qui est per omnia secula benedictus* [who is for all ages blessed].
>
> [31, 1: Appresso questo sonetto [*Oltre la spera che più larga gira*] apparve a me una mirabile visione, ne la quale io vidi cose che mi fecero proporre di non dire più di questa benedetta infino a tanto che io potessi più degnamente trattare di lei. (2) E di venire a ciò io studio quanto posso, sì com'ella sae veracemente; sì che, se piacere sarà di Colui a cui tutte le cose vivono che la mia vita duri per alquanti anni, io spero di dire di lei quello che mai non fue detto d'alcuna; (3) e poi piaccia a Colui che è sire de la cortesia che la mia anima se ne possa gire a vedere la gloria della sua donna, cioè di quella benedetta Beatrice, la quale gloriosamente mira nella faccia di Colui *qui est per omnia secula benedictus. Amen*].

This is the famous "mirabile visione" with which the *Vita Nova* ends, commonly interpreted as a promise and foreshadowing of the composition of a sort of paradisiacal poem, or simply as a promise and foreshadowing, *tout-court*, of the composition of what will be the *Divine Comedy*.

3 Divination by Dreams. *Convivio* II, viii, 7–13 and *Purgatorio* XVII, 13–18

In the second book of the *Convivio* (II, viii, 7–13) Dante puts forward four arguments in support of the immortality of the soul. Here we are interested in the fourth argument, namely the divinatory dreams that make one see the future (II, viii, 13):

> Moreover, we see continual proof of our immortality in the divinations of our dreams, which could not exist if there were not within us some immortal part; since the diviner, whether corporeal or incorporeal, must necessarily be immortal if we stop and consider it (and I say "whether corporeal or incorporeal" based on the diverse opinions

> that I find on the matter), and that which is moved or given form by an unmediated informing agent must be duly proportionate to the informing agent, and there is no proportion between mortal and immortal.
>
> [Ancora: vedemo continua esperienza della nostra immortalitade nelle divinazioni de' nostri sogni, le quali essere non potrebbono se in noi alcuna parte immortale non fosse; con ciò sia cosa che immortale convegna essere lo rivelante, [o corporeo] o incorporeo che sia, se bene si pensa sottilmente – e dico [o] corporeo o incorporeo, per le diverse oppinioni ch'io truovo di ciò –, e quello ch'è mosso o vero informato da informatore immediato debba proporzione avere allo informatore, e dallo mortale allo immortale nulla sia proporzione].

Therefore, says Dante, knowledge of the future can only come from an immortal reality, as such capable of transcending time and having the future as the present. The "different opinions", to which Dante alludes, regarding the corporeal or incorporeal source of divinatory dreams, are found in Albert the Great's *De somno et vigilia*,[3] which presents and criticises the positions of Avicenna, Algazali, Averroes, Alfarabi and Isaac Israeli, supporters of an immediate relationship between angelic intelligences and the human soul (this is the case of the "revelante incorporeo"). Albert the Great, instead, argues in favour of an action of celestial bodies ("revelante corporeo"), which, in Aristotelian physics, despite being material, are not subject to generation and corruption. This is the first passage, in Dante's work, in which an explanation is given of divinatory dreams, and in general of dream visions, a theme for which Dante must have had an acute personal interest, given the intense oneiric and visionary activity of which the *Vita Nova* is an undoubted autobiographical testimony.

Dante then takes up the theme – in exactly the same terms – in the XVII canto of *Purgatory*, to comment on the ecstatic visions that Dante the character is having while walking in a trance along the third and fourth cornice of the mountain of Purgatory, that is the three examples of meekness that he described in *Purg.* XV, 82–128, and the three examples of punished anger that he is preparing to describe now. Before doing so, Dante explains the existence of inner visions, that is, visions that are produced in the mind without coming from the sense of sight:

> O imaginativa che ne rube
> talvolta sì di fuor, ch'om non s'accorge
> perché dintorno suonin mille tube,
>
> chi move te, se 'l senso non ti porge?
> Moveti lume che nel ciel s'informa,
> per sé o per voler che giù lo scorge.
>
> (*Purg.* XV, 13–18)

> [O fantasy, you that at times would snatch / us so from outward things – we notice nothing / although a thousand trumpets sound around us – / who moves you when the senses do not spur you? / A light that finds its form in Heaven moves you – / directly or led downward by God's will].

But what is the source, or what are the sources, of this explanation, which appears for the first time in the second book of the *Convivio*, the second time in the XVII canto of *Purgatory*? The still fundamental study on this point today, that is, the 1938 essay by Bruno Nardi, enumerates a series of Arab, Jewish, and Latin commentators on Aristotle (Avicenna, Algazel, Moses Maimonides, and Pietro d'Abano) as remote sources,[4] and Albert the Great as Dante's most likely immediate source. Precisely, Nardi refers to a passage from *De natura et origine animae* in which Albert summarises an argument on the immortality of the soul held by ancient sages. The argument, in turn, was confirmed in Aristotle's *On Sleeping and Sleeplessness*:

> And this reason [*i.e.* the previous argument for the immortality of the soul] is confirmed by means of another, that is taken from the sayings of Socrates and Hermes Trismegistus [...]. And the reason for this [*i.e.* the rightness of the latter argument] is that, as it has been proven in *On Sleeping and Sleeplessness*, there are certain true intellections (*intelligentiae*) in dreams that do not have the nature of a dream but rather are oracles flowing (*influxa*) by the higher intellects (*ab intellectibus supernis*) into the intellect of the soul (*in intellectum animae*). However, it is established that their reception is only according to the conformity of the human soul to the higher and celestial intellects, and not according to any dependence on the body. Therefore, such a substance that receives these oracles cannot perish with the destruction of the body.[5]

In fact, this short passage by Albert the Great contains the essence of the idea expressed by Dante. The core of this idea seems to date back to Avicenna, but it is widely believed that Dante did not know Avicenna's *De anima* directly, but only through Albert the Great, particularly the chapters of *De somno et vigilia* mentioned above, and Thomas Aquinas.[6] However, Avicenna's text had been translated into Latin by Gundissalinus already between 1125 and 1150, even before the works of Aristotle, and it was widespread in the 13th century.[7]

In this case, therefore, I would exclude that Dante formed his belief only on Albert the Great's *De somno et vigilia*, because this treatise is profoundly doubtful about the existence and validity of *divination*.[8] Instead, Dante takes the existence of the "divinations of our dreams" as a fact, of which he says that we have "continuous experience". Indeed, a close and articulated correspondence between the thoughts expressed by Dante and Avicenna emerges clearly from book 4, chapter 2 of Avicenna's *Liber de anima seu sextus de naturalibus*, entitled "On the actions of these inner senses, namely the formal and cogitative, in which the discussion pertains to sleep and wakefulness, deceptive and true dreams, and a little about the nature of prophesying"

("De actionibus horum sensuum interiorum, formalis scilicet et cogitativae, in quo tractatur de somno et vigilantis et de somno fallaci et vero et aliquantulum de proprietate prophetandi"), and from book 5. In fact, they present rather binding concordances (although none of them are exclusive) with various passages by Dante, and a couple of statements that seem to reveal a notable consonance with Dante's personality and his dream-visionary history.

Here are these concordances:

A The concept – as in the passage just quoted from Albert's *De natura et origine animae* – that prescient dreams demonstrate the affinity between the human soul and angelic intelligences:

> From the discussion of imagination, we now come to dreams. We should consider a little of how something comes to be foreseen in dreams from those things that are deposited there. A fuller account is given in the works dealing with metaphysics. Everything in the world, past, present and future, exists in the wisdom of the creator and angelic understanding according to some, and in the souls that are the angels of the heavens according to others. Later <in the metaphysical work> I will enter into this in more detail and show that human souls have a stronger relationship to angelic substance than to the physical bodies. Obscurity [of prophetic and meaningful dreams] does not come about from lack of interest, but from how the information is received. For either it is infused on physical bodies or it is polluted by circumstances that degrade performance.[9]

B The notion that prescient dreams occur at dawn:

> Moreover, dreams are often not true unless they occur in the morning. All cogitation ceases at this time and the movements of the humours are completed. At this time, the body does not impede the imaginative power, and it can communicate with the memorial power and the image store. Then, the service of the imagination that serves the soul is as good as it can be. This is important because it is necessary that the form that arrives there is clearly depicted in these powers, either as it is or through its representation.[10]

Certainly not an exclusive notion, but recurring in the *Comedy* and especially functional to its narrative-visionary structure. Suffice it to remember *Inf.* XXVI, 7: "Ma se presso al mattin del ver si sogna…" ("But if the dreams dreamt close to dawn are true…"), and the three dreams of *Purgatory*, all three morning dreams, introduced by the anaphora "Ne l'ora che", that is:

> Ne l'ora che comincia i tristi lai
> la rondinella presso a la mattina,
> forse a memoria de' suo' primi guai,

e che la mente nostra, peregrina
più da la carne e men da' pensier presa,
a le sue vision quasi è divina,

in sogno mi parea veder sospesa
un'aguglia nel ciel con penne d'oro,
con l'ali aperte e a calare intesa.
(*Purg.* IX, 13–21)

[At that hour close to morning when the swallow / begins her melancholy songs, / perhaps in memory of her ancient sufferings, / when, free to wander farther from the flesh / and less held fast by cares, our intellect's / envisionings become almost divine – / in dream I seemed to see an eagle poised, / with golden pinions, in the sky: its wings / were open; it was ready to swoop down].

Ne l'ora che non può 'l calor diurno
intepidar più 'l freddo de la luna,
vinto da terra, e talor da Saturno

quando i geomanti lor Maggior Fortuna
veggiono in oriente, innanzi a l'alba,
surger per via che poco le sta bruna,

mi venne in sogno una femmina balba
(*Purg.* XIX, 1–7)

[In that hour when the heat of day, defeated / by Earth and, sometimes, Saturn, can no longer / warm up the moon – sent cold, when geomancers / can, in the east, see their Fortuna major / rising before the dawn along a path / that will be darkened for it only briefly – / a stammering woman came to me in dream].

Ne l'ora, credo, che de l'oriente,
prima raggiò nel monte Citerea,
che di foco d'amor par sempre ardente,

giovane e bella in sogno mi parea
donna vedere andar per una landa
(*Purg.* XXVII, 94–98)

[It was the hour, I think, when Cytherea, / who always seems aflame with fires of love, / first shines upon the mountains from the east, / that, in my dream, I seemed to see a woman / both young and fair; along a plain].

C The notion that the soul, when it is entirely occupied in one faculty, makes the others inactive, a fact that Avicenna, like Dante, brings as proof of the uniqueness of the soul:

> To make this statement clearer, in what follows we emphasise that all the powers are powers of one single soul and serve it. If you bear this in mind, you will realise that, with respect to each of these powers, the commitment of the soul to preserve some of these [powers], restrains it [the soul] from helping the others, from keeping them out from error, and from leading them on the correct path. However, when the soul is busy with interior matters, it is not able to attend to external matters as much as it should. When it is busy with external matters, it neglects to govern the interior powers. When it considers external objects intently, its imagination and memory weaken. When it focusses on the actions of the desiring power, the actions of the power of arousal weaken; and when it focusses on the actions of the power of arousal, the actions of the desiring power weaken. Similarly, when it focusses on such motivating actions, its actions of perception and apprehension weaken, and vice versa.[11]

This is a recurring notion in the *Comedy*:

> Quando per dilettanze o ver per doglie
> che alcuna virtù nostra comprenda
> l'anima bene ad essa si raccoglie,
>
> par ch'a nulla potenza più intenda;
> e questo è contra quello error che crede
> ch' un'anima sovr'altra in noi s'accenda.
>
> E però, quando s'ode cosa o vede
> che tegna forte a sé l'anima volta,
> vassene 'l tempo e l'uom non se n'avvede;
>
> ch'altra potenza è quella che l'ascolta,
> e altra è quella c'ha l'anima intera:
> questa è quasi legata, e quella è sciolta.
>
> Di ciò ebb'io esperienza vera,
> udendo quello spirto e ammirando;
> ché ben cinquanta gradi salito era
>
> lo sole, e io non m'era accorto, quando
> venimmo ove quell'anime ad una
> gridaro a noi: "Qui è vostro dimando".
> (*Purg.* IV, 1–18)

[When any of our faculties retains / a strong impression of delight or pain, / the soul will wholly concentrate on that, / neglecting any other power it has / (and this refutes the error that maintains / that – one above the other – several souls / can flame in us); and thus, when something seen / or heard secures the soul in stringent grip, / time moves and yet we do not notice it. / The power that perceives the course of time / is not the power that captures all the mind;/ the former has no force – the latter binds. / And I confirmed this by experience, / hearing that spirit in my wonderment; / for though the sun had fully climbed fifty / degrees, I had not noticed it, when we / came to the point at which in unison / those souls cried out to us: "Here's what you want"].

And again:

Tant'eran li occhi miei fissi e attenti
a disbramarsi la decenne sete,
che li altri sensi m'eran tutti spenti.

Ed essi quinci e quindi avien parete
di non caler – così lo santo riso
a sé traéli con l'antica rete! –;

quando per forza mi fu vòlto il viso
ver' la sinistra mia da quelle dee,
perch'io udi' da loro un "Troppo fiso!";

e la disposizion ch'a veder èe
ne li occhi pur testé dal sol percossi,
sanza la vista alquanto esser mi fée.
(*Purg.* XXXII, 1–12)

[My eyes were so insistent, so intent / on finding satisfaction for their ten-/year thirst that every other sense was spent. / And to each side, my eyes were walled in by / indifference to all else (with its old net, / the holy smile so drew them to itself), / when I was forced to turn my eyes leftward / by those three goddesses because I heard / them warning me: "You stare too fixedly". / And the condition that afflicts the sight / when eyes have just been struck by the sun's force / left me without my vision for a time].

D The notion that the imagination produces internal images identical to those perceived by the senses when it is asleep, or debilitated by infirmity (e.g. epilepsy), or by terror.

After stating that the soul can prevail on imagination in one of the two ways, that is either by subjecting it to its will in a conscient way, that is through intellect and reason, or by recognising that the appearances contrived by imagination do not correspond to the external objects, Avicenna says:

> If the imaginative power is restricted in either way, its action is limited. The restrictions may be removed, as happens when people are asleep. This may also occur because of disorders that debilitate the body and prevent the soul from having proper use of intellect and cognition. Also, when people are terrified, the soul is disabled, and it is as if it disintegrates from fear and retreats completely from the intellect owing to this disability, and because it fears bodily harm. So the soul may withdraw from the intellect and ruling through it, and then imagination can exert itself and turn to the image store (*ad formalem*) and command it to operate, and the bond between them is strengthened. The action of the image store becomes clearer, and forms in it appear in the common sense and they seem to exist externally. The impression perceived (*operatio apprehensa*) from what comes from the external world and what comes from the interior senses is what is presented in the image store (*in formali*), and the difference is only in the source. But when a thing is truly sensed as appearing in it [i.e. in the imaginative power], then what comes from the inside is seen as though coming from outside. In this way, epileptic, terrified, dissolute, or sleepy people see imagined things as existing with those qualities that they see externally when they are healthy. Similar misapprehensions can occur with sounds.[12]

A passage to be compared, in particular, with the episode of the hallucination due to the "dolorosa infirmitade" of the *Vita Nova* (above in Section 2.5).

E The notion that the *imaginativa* produces internal images identical to those perceived by the senses when it is very strong, so much so that it is not dominated by the senses. This happens when the soul of the individual is very strong, very noble, and intellectual contemplation imposes itself on the senses. Individuals with this strength experience visions while awake, not just in dreams. This is precisely the prophecy produced by the *imaginativa*:

> Certain people have a very strong imaginative power. Their senses do not dominate it and the image store (*formalis*) does not resist it. Their souls are very strong, and hence, when the soul surveys the intellect and what is above the intellect, its ability to rule the senses still survives. These people have capabilities when awake that others have in dreams, as we see later in this chapter. This is the state of sleeping people when they perceive visions, so that they are informed either about how things really are or through images that they have. These people also have such things happen when awake. Sometimes they will be between the waking and sleeping states and very much absent from the sensory

> world, just as if they were asleep. Sometimes this does not happen, and they see something just as it is, and often an image appears and it is imagined while sleeping, as we shall see later in this chapter. Sometimes it seems to them that what they perceive is a speech made by an image, like audible words that they hold and read. This is the prophetic property of the imaginative power.[13]
>
> Of the people who see these things while awake, some do so because of the nobility of their souls, because their souls are strong, and because of the strength of their imaginations and memories, whose actions are not impeded by external distractions.[14]

I believe that these last two passages from Avicenna's *De anima* were particularly important for Dante when he wrote the second book of the *Convivio*. Indeed, in his philosophical research, when he focuses on what dreams and visions are, Dante has behind him all the dream-visionary experience lived and narrated in the *Vita Nova*, all the work of his imagination that for the entire duration of his association with Beatrice created a counter-script of internal facts intertwined with the few external facts occurring in the meanwhile. These internal facts were precisely dreams, fantasies produced by the too strong emotion induced by the presence of Beatrice, hallucinations caused by some infirmity. Therefore, a case study very similar to that just described by Avicenna, who furthermore considers such manifestations a sign not of weakness but of inner strength, of nobility of the soul.

I believe that this intellectual discovery by Dante was decisive in giving great authority in his eyes to his dream-visionary inclination, and thus encouraging him to launch himself into writing the poem that would give shape to the total vision of the three kingdoms of the afterlife.

Notes

1 The contents summarised in this article are covered more extensively in Tavoni, "Dante 'Imagining' His Journey"; "L'Inferno sognato, la telepatia di Virgilio" and "La visione interiore dalla *Vita nova* al *Convivio*".

2 The *Vita Nova* is cited in Italian from the Gorni edition, in English from the Cervigni-Vasta edition (which follows the numbering system of the Barbi edition). References to Barbi edition are provided within brackets, preceded by those of Gorni edition. The *Convivio* is cited in Italian from the Fioravanti edition, in English from the Frisardi edition. The *Comedy* is cited in Italian from the Petrocchi edition, in English from the translation by Allen Mandelbaum in *Digital Dante* (https://digitaldante.columbia.edu/).

3 Albert the Great, *De somno et vigilia* III, treatise 1, ch. 5–12.

4 Nardi, "L'immortalità dell'anima": 231–238 and "Il sogno divinatòrîo".

5 Albert the Great, *Liber de natura et origine animae* (treatise 2, ch. 6), 28 lines 35–50: "Confirmatur autem haec ratio per aliam, quae sumitur a dictis Socratis et Hermetis Termegisti [...] Et fuit ratio eorum, quod, sicut in *Somno et vigilia* probatum est, quaedam sunt in somnis verae intelligentiae, quae non somnii habent rationem, sed potius oracula sunt, ab intellectibus supernis in intellectum animae influxa. Constat autem, quod horum receptio non est nisi secundum

conformitatem animae humanae ad intellectus supernos et caelestes et non secundum aliquam dependentiam ad corpus, et ideo nec perire potest huiusmodi substantia talia recipiens oracula corpore pereunte". My translation.

6 Giacon, "Avicenna", 481: "è opinione comune [...] che Dante ha conosciuto il pensiero di Avicenna indirettamente, specialmente attraverso Alberto Magno e Tommaso d'Aquino"; but at 482: "Un'altra reminiscenza può essere la *divinatio in somniis* come prova razionale per l'immortalità dell'anima", referring precisely to *Convivio* II, viii, 13. On the medieval reception of ideas about dreams, stars and prophecy see Gregory, "I sogni e gli Astri", Grellard, "La réception médiévale".

7 On the diffusion and importance of Avicenna's *De anima* in the 13th century see Hasse, *Avicenna's* De Anima: 154–174 and Verbeke, *L'immortalité de l'âme.*

8 Albert the Great, *De somno et vigilia* III, treatise 1, "Quod sit aliqua divinatio et quid sit", in particular ch. 2 ("An divinatio somniorum sit, an non sit"), 179–180.

9 Avicenna, *Liber de anima* V, 2, 28–29: "Sed quia de verbo imaginationis iam pervenimus ad somnia, bonum est ut ostendamus parum principium ex quo provenit aliquid praevidere in somnis ex his quae ponemus; non enim firmiter monstrabuntur nisi in philosophia prima. Dicemus ergo quod omnia quae in mundo sunt praeterita, praesentia et futura, habent esse in sapientia creatoris et angelorum intellectualium secundum aliquid, et habent esse in animabus quae sunt angeli caelorum secundum aliud. Postea autem declarabuntur isti duo modi alias et demonstrabitur tibi quod animae humanae maiorem habent comparationem cum substantiis angelicis quam cum corporibus sensibilibus, et non est illic occultatio aliqua nec avaritia, sed occultatio est secundum receptibilia, aut quia sunt infusa corporibus, aut quia sunt inquinata ab his quibus deprimuntur deorsum". This and the following translations reproduce, with some variations, the translation by Simon Kemp, *The Psychology of Avicenna.* This online translation has no page numbering.

10 Avicenna, *Liber de anima* V, 2, 32: "Praeter hoc etiam non sunt vera somnia plerumque nisi quae videntur in mane: omnes enim cogitationes hac hora quiescunt et motus humorum sunt finiti; cum vero virtus imaginativa fuerit in tali hora qua non impeditur propter corpus nec est separata a memoriali nec a formali, sed est compos illarum, tunc servitium imaginativae quo servit animae est quale melius esse potest: necesse est enim sine dubio ut forma quae advenit illi, describatur in istis virtutibus firmiter, aut ita ut est aut per conformationem".

11 Avicenna, *Liber de anima* IV, 2, 13–14: "Ad maiorem autem huius declarationem dicimus nos in sequentibus demonstraturos quod omnes hae virtutes virtutes sunt unius animae deservientes ei. Crede hoc nunc sic positum. et scias quod occupatio animae circa aliquam istarum retinet eam ne adiuvet alias et ne conservet eas ab errore ipsarum et ducat eas ad viam rectitudinis. Anima etenim cum occupata fuerit circa interiora, non solet curare de exterioribus quantum deberet; et cum occupata fuerit circa exteriora, praetermittet gubernare virtutes interiores. Ipsa enim cum intente considerat sensibilia exterius, ea hora qua de his tractat, debilitatur eius imaginatio et memoria; cum vero oboedit actionibus virtutis concupiscibilis, debilitantur actiones virtutis irascibilis; sed cum oboedierit actionibus virtutis irascibilis, debilitantur actiones virtutis concupiscibilis; et omnino cum oboedierit dispositioni actionum motivarum, debilitabuntur actiones apprehensivae et e converso".

12 Avicenna, *Liber de anima* IV, 2, 17–18: "Si autem imaginativa impedita fuerit utroque modo, debilitabitur eius actio. Sed remoto utroque impedimento sicut fit in hora dormiendi, aut ex uno modo – sicut in infirmitatibus quae debilitant corpus et impediunt animam ne habeat intellectum et cognitionem, et sicut fit in terrore cum debilitatur anima et quasi dissolvitur propter id quod timetur et retrahitur omnino ab intellectu propter suam debilitatem et quia timet ne accidant res corporales, et sic quasi discedit ab intellectu et ab eius dominio, – tunc imaginatio potest niti et converti ad formalem et iniungere ei operari, et earum adunatio

confortatur simul, et actio formalis fit manifestior, et formae quae sunt in formali praesentantur in sensu communi et videntur quasi habeant esse extrinsecus: operatio etenim apprehensa ex eo quod venit ab exterioribus et ex eo quod venit ab interioribus est id quod praesentatur in formali, et non differunt nisi comparatione. Sed cum sensatum vere est id quod apparet in ea, tunc id quod apparet in ea intus est tale quale est quod apparet foris; et propter hoc videt epilepticus et perterritus et dissolutus et soporatus imaginationes existentes qualiter vere videt in tempore salutis et etiam audit sonos".

13 Avicenna, *Liber de anima* IV, 4, 18–19: "Contingit autem aliquibus hominibus quod haec virtus imaginativa sit creata in illis fortissima et praevalens, ita ut non dominentur ei sensus nec formalis resistat ei, et quod anima eorum sit fortissima, ita quod propter hoc quod contemplatur intellectum et id quod est supra intellectum, non destruatur eius condescensio ad sensus. Isti habent in vigiliis quod alii in somnia, sicut postea dicemus: haec enim est dispositio dormientis dum apprehendit visiones, ut certificentur ei aut ita ut sunt, aut per imagines quas habent; istis quoque accidunt talia in vigiliis. Saepe enim inter utrumque istorum contingit eos in ultimo absentari a sensibilibus et accidit eis quasi dormitatio; et multotiens non accidit, et multotiens vident rem sicuti est, et multotiens apparet eius imago propter causam enim qua imaginatur in dormiente imago rei quae videtur, sicut postea declarabimus; multotiens apparet similitudo et videtur eis quod id quod apprehendunt sit locutio illius imaginis veluti verba audita quae tenent et legunt. Et haec est propria prophetia virtutis imaginativae".

14 Avicenna, *Liber de anima* IV, 4, 26: "Eorum autem qui haec vident vigilantes, quidam vident sic propter nobilitatem suae animae et quia eius anima fortis est et propter fortitudinem suae imaginationis et memoriae, quam non impediunt sensibilia a suis propriis actionibus".

Bibliography

Primary Sources

Albert the Great. *De somno et vigilia*. Edited by Auguste Borgnet. Paris: L. Vivès, 1890.

Albert the Great. *Liber de natura et origine animae*. Edited by Bernhard Geyer. Münster: Aschendorff, 1955.

Avicenna. *Liber de anima seu sextus de naturalibus*. Edited by Simone Van Riet, 2 vols. Louvain and Leiden: Éd. Orientalistes-E.J. Brill, 1968–1972.

Avicenna. *Liber de anima*. English translation by Simon Kemp. *The Psychology of Avicenna: An English Version of the Liber de Anima*. Cambridge: UC Library, 2023. https://hdl.handle.net/10092/106061.

Dante Alighieri. *La vita nuova*. Edited by Michele Barbi. Florence: Bemporad, 1932.

Dante Alighieri. *La Commedia secondo l'antica vulgata*. Edited by Giorgio Petrocchi, 4 vols. [Second revised edition]. Florence: Le Lettere, 1994.

Dante Alighieri. *Vita Nuova*. [Italian text with facing] English translation by Dino S. Cervigni and Edward Vasta. Notre Dame: The University of Notre Dame Press, 1995.

Dante Alighieri. *Vita Nova*. Edited by Guglielmo Gorni. Milan: Mondadori, 2011.

Dante Alighieri. *Il Convivio*. Edited by Gianfranco Fioravanti. Milan: Mondadori, 2014.

Dante Alighieri. *Convivio. A Dual-Language Critical Edition*. Edited and translated by Andrew Frisardi. Cambridge: Cambridge University Press, 2018.

Secondary Sources

Digital Dante. Original Research & Ideas. The Divine Comedy with Commento Baroliniano. https://digitaldante.columbia.edu/.

Giacon, Carlo. "Avicenna". In *Enciclopedia Dantesca,* edited by Istituto della Enciclopedia Italiana, vol. 1, 481–482. Roma: Istituto della Enciclopedia Italiana, 19842.

Gregory, Tullio. "I sogni e gli astri". In *I sogni nel medioevo*, edited by Tullio Gregory, 111–148. Roma: Edizione dell'Ateneo, 1985.

Grellard, Christophe. "La réception médiévale du *De somno et vigilia.* Approche anthropologique et épistémologique du rêve, d'Albert le grand à Jean Buridan". In *Les Parva naturalia d'Aristote. Fortune antique et médiévale*, edited by Pierre-Marie Morel and Christophe Grellard, 221–237. Paris: Editions de la Sorbonne, 2010.

Hasse, Dag Nikolaus. *Avicenna's De Anima in the Latin West: The Formation of a Peripatetic Philosophy of the Soul, 1160–1300.* London and Turin: The Warburg Institute-Nino Aragno Editore, 2000.

Nardi, Bruno. "Il sogno divinatòrîo". In *Saggi di filosofia dantesca*, edited by Bruno Nardi, 55–58. Florence: La Nuova Italia, 1967.

Nardi, Bruno. "L'immortalità dell'anima". In *Dante e la cultura medievale*, edited by Paolo Mazzantini, 225–243 [first edition 1938]. Roma and Bari: Laterza, 1990.

Tavoni, Mirko. "Dante 'Imagining' His Journey through the Afterlife". *Dante Studies* 133 (2015): 70–97.

Tavoni, Mirko. "L'Inferno sognato, la telepatia di Virgilio e gli antefatti danteschi della *Commedia* come visione in sogno". In *Dante e la dimensione visionaria tra medioevo e prima età moderna. Atti del Seminario di studio, Freie Universität Berlin, 23 novembre 2017,* 97–119, edited by Bernhard Huss and Mirko Tavoni. Ravenna: Longo, 2019.

Tavoni, Mirko. "La visione interiore dalla *Vita Nova* al *Convivio*". In *"Vedi lo sol che 'n fronte ti riluce". La vista e gli altri sensi in Dante e nella ricezione artistico-letteraria delle sue opera*, edited by Maria Maślanka-Soro and Anna Pifko-Wadowska, 43–66. Rome: Aracne, 2019.

Verbeke, Gérard. *L'immortalité de l'âme dans le De Anima d'Avicenne. Une synthese de l'aristotélisme et du néoplatonisme.* Madrid: Centro Loyola de Estudios y Comunicacion Social, 1969.

6 Seeing the Light

Dante and the Perspectivist Theory of Light as Proper Visible

Cecilia Panti

1 Introduction: Which Theory of Light Did Dante Follow?

As Mark Smith stresses in his magisterial book *From Sight to Light*, up to modern age "the science of optics was aimed primarily at explaining not light and its physical manifestations, but the sight in all its aspects from physical and psychological causes to perceptual and cognitive effects".[1] This fact, as Smith remarks, had a noticeable consequence on the Western ancient and medieval conception of light, namely that "light theory was not only regarded as subsidiary to sight theory but was actually accommodated to it".[2] This is what happens in the two main sight theories diffused at Dante's time, namely the Graeco-Roman (and Arab) geometrical and medical model of the emission of visual rays or spirits from the eye and the Aristotelian model of the intromission of forms. The former model developed in Platonic and Neoplatonic sources, and in the scientific and medical writings embracing pseudo-Euclid, Ptolemy, Galen, Alkindi, and Alfarabi. The latter model, instead, had divulged in the Latinate world since the beginning of the 13th century with the entrance of Aristotle and his Arab commentators, mainly Avicenna and Averroes, and was discussed and adopted in the 13th-century Latin commentaries, including those by Albert the Great and Thomas Aquinas. In both models, light is not an object of sight, but what makes colours visible to sight.

In Dante's time, however, a third theory of vision had started diffusing, namely that by the Arab scientist Ibn al-Hytham (Alhazen, or, in the common Latin spelling, Alhacen), exposed in his *Book of Optics* (Kitab al-Manazir, c1030). This optical model introduced the principle that sight perceives primarily irradiated light, an idea that allowed Alhazen to geometrise the intromissionist theory of the reception of forms. Light, or better light ray, is, therefore, a proper visible, together with – but independent from – colour. Therefore, while previous theories "treated light as a catalytic agent […] to render color visible, not to be seen in its own right", for Alhazen "light was per se visible, seen in its own right rather than through its effect on color. The contrast between these two conceptions of light could hardly be clearer – at least in principle".[3]

DOI: 10.4324/9781003412601-7

Alhazen's masterpiece was translated into Latin at the beginning of the 13th century, under the title *De aspectibus* and, later, *Perspectiva*. Its reception begins in the middle of the century with the rising of scientific writings on *perspectiva*, a term first adopted by Robert Grosseteste to qualify the extramissionist geometrical optics and its function in the explanation of the rainbow.[4] This same term was promptly assumed to qualify also Alhazen's geometrical model of vision.[5] The "perspectivists" (*perspectivi*), consequently, embraced a geometrical approach to the science of vision, independently of their intromissionist or extramissionist model.[6]

The first treatises to bear the title *Perspectiva* are works on optics by two Franciscan masters: Roger Bacon and John Peckham (or Pecham). Bacon's *Perspectiva*, written in the '60s was sent to Pope Clement IV in 1267.[7] Shortly after, at the papal court of Viterbo, Peckham wrote the *Perspectiva communis*, which promptly became the most diffused perspectivist synthesis, and the *Tractatus de perspectiva*, where geometrical optics is used as a source of theological metaphors. Again, at Viterbo, early in the '80s, appears the *Perspectiva* by the Silesian scientist and philosopher Witelo, who offered a systematic geometrical explanation of Alhazen's optics and theory of light to the Latinate scholars.[8] In these works, the idea that the light ray is an object of sight is strongly stated, and, in the case of Bacon, even 'accommodated' within a partially extramissionist approach.

What follows aims to show that Dante adopted the new central idea of perspectivism concerning light as proper visible, even in the *Vita Nuova* and the early *Rime*. Moreover, contrary to what is generally held, this paper also claims that Dante did not adopt the extramissionist model in these compositions.[9] Let us start with the latter issue.

2 The Alleged Extramissionist Theory in the *Vita Nuova*

The optical theory commonly called extramissionist, or extramissive, was unanimously followed in the early Middle Ages and had important supporters also in the late Middle Ages, and even later. It was a highly elaborated scientific theory, based on the Platonic, Euclidean, Ptolemaic, and Galenic science of vision, which spread in the Latinate world also thanks to the Arab optical literature. Notwithstanding its noticeable variants, this theory asserted that vision is due to the emission of flux or radiation from the eye, generally called in Latin *radius visualis* (visual ray) or also *spiritus visus* or *visualis* (visual spirit). This theory of vision had its geometrical foundation in the Euclidean assumption that the eye emits radial lines bundled together to form a cone with its vertex in the eye and the basis on a given object. Ptolemy improved this model with the idea of a continuum visual flux, similar to the Stoics' pneumatic stream, forming the visual cone; moreover, he suggested that the flux senses colours and the object's characteristics only if the object is compact and illuminated. Luminosity as such is not properly visible but grounds the visibility of colour, which is sensed by vision. Even luminous objects are not seen because of their luminosity but, rather, thanks to their bright colour. Similarly, Galen postulated that the visual pneuma passes through the hollow optic nerves into the eyes, flowing through the pupil into the surrounding air. If air is illuminated, the pneuma transforms it into the instrument through which the eye 'touches' external objects. Again, for Galen what is sensed is colour, through which the pneuma perceives other objectual properties such as size and shape.

These ancient theories had a great impact on Arabic optics. The Persian scientist Alkindi (d. 873) merged them with his theory of irradiated light, describing the visual cone as a luminous power issued from the eye. He also developed two pivotal concepts, widely diffused among Latins, namely that light propagates according to rectilinear lines everywhere and instantaneously and that visual acuity along the axis of the visual cone is stronger.[10] Alfarabi (d. 950), in turn, introduced this mathematical approach to optics in his interpretation of Aristotle's theory of vision.[11]

Western scholastic thinkers knew the extramissionist model in a sort of abridged and eclectic form based on the mentioned theories. For example, Albert the Great's *De Sensu* refutes the extramissionist stance with long and convoluted arguments recounting the theories of vision of Plato, Galen, and Alfarabi.[12] Dante, instead, simply states in *Conv.* III, ix, 10 that extramissionism was proved to be false by Aristotle:

> Plato and other philosophers, however, said that our sight was not a result of the visible entering the eye but of the visual power going out to the visible, but this opinion is rejected as false by the Philosopher in his book *On Sense and Sensibles*.[13]

Several Dante scholars believe that extramissionism was in fact the theory of vision followed by Dante in his early poetic compositions.[14] Moreover, it has been also assumed that Dante deliberately adhered to this theory even in the *Divine Comedy*: since the hell is a place devoid of light, Dante could see in the darkness only by means of his visual rays.[15] In my opinion, not only is this idea absent from the *Commedia*, but it is not at all present in Dante, even in his early writings.

The passages in the *Vita Nuova* and *Rime* commonly adduced to support Dante's theory of the visual rays are those concerning "spirits" (*spiriti*) or "light" (*lume*) or "rays" (*raggi*) departing from the eye.[16] It has been assumed that what departs from the eye is, *ipso facto*, the sensitive instrument, namely the visual spirits. Hereafter, I challenge this assumption by considering four passages (A to D) dealing with Beatrice's eyes and another two (E and F) concerning Dante's eyes.

> A. *Vita Nuova* XIX, 12: "De li occhi suoi, come ch'ella li mova, / escono spirti d'amore inflammati, / che fèron li occhi a qual che allor la guati; /

e passan sì che 'l cor ciascun retrova: / voi le vedete Amor pinto nel viso, / là 've non pote alcun mirarla fiso".

[her eyes, wherever she may choose to move them, send forth their spirits radiant with love to strike the eyes of anyone looking at her, and penetrate until they find the heart. You will see Love depicted on her face, there where no one dares hold his gaze too long].

B. Rima LXIX: "De gli occhi suoi gittava una lumera, / la qual parea un spirito infiammato".

[From her eyes she cast a light that seemed like a spirit aflame].

C. Rima LXV: "Degli occhi della mia donna si move / un lume sì gentil, che dove appare / si veggion cose ch'om non po' ritrare".

[From my lady's eyes there comes a light so noble that where it appears things are seen that no one can describe].

D. Rima XCI: "Entrano i raggi di questi occhi belli / ne' miei innamorati".

[The rays from these fair eyes enter my own loving eyes].

Passage *A* asserts that "spirits inflamed with love" exit from Beatrice's eyes and wound the poet's eyes. If Dante were following, here, the extramissionist stance, he himself would be the object of sight for Beatrice. But is Beatrice looking at Dante? Are Beatrice's spirits her visual spirits, namely the ones that, in reaching Dante, convey his image to the woman, enabling her to see him? Clearly, the answer is no. In fact, these "spirits inflamed with love" are rather rays bringing love and piercing Dante's eyes to go straight to his heart. These spirits are the reason why the poet cannot look at Beatrice's eyes, for they are like "light causing pain" to him, as the same verse clarifies. In passage *A*, therefore, there is no allusion to the extramissionist thesis; rather, to the intromissionist model, and specifically to the perspectivist thesis that something "fiery" and luminous exits from the observed object (here, Beatrice's eyes) and enters the observer's eyes. The passages referred to in *B*, *C*, and *D* clarify what Dante means by "spirit of love" or "inflamed spirit". It is a sort of irradiated light (*lumera/lume*) and "rays that wound the eyes" of the poet. Dante, therefore, is not stating that Beatrice sees thanks to the light/spirits/rays emitted from her eyes, but exactly the opposite: namely, that Dante sees the light emitted from Beatrice's eyes, a light that enters his eyes, as stated in *D*, causing him extreme pain before passing on to his heart.

But let us now consider what happens in Dante's eyes. Again, he refers to a displacement of spirits including, now, also the "visual spirits" ("spiriti del viso"), which are "pushed away of their place":

> *E. Vita Nuova* XI, 2: "And when she was about to greet me, one of Love's spirits, annihilating all the others of the senses (*spiriti sensitivi*), would drive out (*pingea fuori*) the feeble spirits of sight (*deboletti spiriti del viso*), saying to them, 'Go and pay homage to your mistress', and Love would take their place (*si rimanea nel loro luogo*). And if anyone had wished to know Love, he might have done so by looking at my glistening eyes".
>
> *F. Vita Nuova* XIV, XII, 7: "ché Amor, quando sì presso a voi mi trova, / prende baldanza e tanta sicurtate, / che fere tra' miei spiriti paurosi, / e quale ancide e qual pinge di fore, / sì che solo rimane a veder voi".
>
> [for Love, when he discovers me near you, takes on a cruel, bold new confidence and puts my frightened senses to the sword, by slaying this one, driving that one out, till only he is left to look at you].

Passages *E* and *F* emphatically underscore that love destroys Dante's sensitive spirits and pushes his visual spirits out of their abode, where love positions itself. Love, then, invites the visual spirits to "pay homage" to Beatrice, which would mean – as commonly held – to go near her for looking at her. However, passage *F* clarifies that the displaced spirits cannot do it anymore, since love is now looking at Beatrice. Noticeably, these sentences state that sight requires that visual spirits *be and remain* in their abode for allowing vision. This same idea, that both passages remark, is fully formulated in *Vita Nuova* XIV, 4–6, where Dante repeats what happens to his spirits and visual spirits when looking at Beatrice:

> I raised my eyes and, looking at the ladies, I saw among them the most gracious Beatrice. Then my spirits were so disrupted by the strength Love acquired when he saw himself this close to the most gracious lady, that none survived except the spirits of sight (*spiriti del viso*); and even these were driven forth of their instruments (*rimasero fuori de li loro istrumenti*) because Love desired to occupy their enviable post (*nel loro nobilissimo luogo*) in order to behold (*per vedere*) the marvelous lady. And even though I was not quite myself, I was still very sorry for these little spirits who bitterly protested, saying: 'If this one had not thrust us from our place like a bolt of lightning, we could have stayed to see (*se non ci infolgorasse così fuori del nostro luogo, noi potremmo stare a vedere*) the wonders of this lady as all our peers are doing'.[17]

The dislocation of Dante's visual spirits is an effect of the extraordinary inner devastation due to the exceptional nature of the object seen, Beatrice. Vision, in fact, occurs *only if* the spirits stay in their place, as Dante expressly asserts here by reporting the spirits' complain: "outside of our place, we cannot see Beatrice". Indeed, love alone, which enthroned itself in their place, can now admire such an outstanding *beatitudine*. But where is the visual spirits' place? The passage is clear: it is in the *instrumenti*, namely the optic nerve, as Dante also asserts in *Conv.* II, IX, 5: "the nerve through which the visual spirit runs"; a nerve that starts in the brain chiasm and terminates in the eyes.

In conclusion, the emanation of light or radiance or spirit of love from the eyes of the beloved woman – a ubiquitous theme in Dante, as well as in the *stilnovo* poetry – does not mean, at least for Dante, the flux of visual spirits reaching the object, as the extramissionist model implies. This emanation, instead, is an object of sight for Dante; it is, exactly, the irradiated light that the perspectivists conceived of as something visible in its own right. As we see in the next two sections, Dante thought that also colour and the other objectual forms are received into the eye according to straight lines, the same lines through which irradiated light is received. Dante, therefore, follows an intromissionist theory of vision. Hereafter, we will see that it is not the Aristotelian intromissionism, but a coherent and very simplified perspectivism inspired by Alhazen's optics.

3 Aristotelian Intromissionism and Theory of Light

The intromissionist model of vision is described in *Convivio* III, ix, 6 with open reference to Aristotle's *On the Soul* book 2 and *On Sense*, as we shall see in the next section of this study, where this reference will be considered in the light of an error of punctuation that has seriously misled its interpretation. To address this issue, let us first recapitulate the theory of light and vision of Aristotle and his commentators.

For Aristotle, sight depends upon the uninterrupted physical contact between the eye and external objects due to the air or water staying in between.[18] Air or water, however, are only potentially "media" for sight; in fact, they are so only if their transparency is "in act". This happens in the presence of a source of light, that is, the sun, fire, or the moon. Thanks to this qualitative transformation of the medium, the object's colour is instantaneously imprinted on the eye, like a seal in soft wax. From colour impression, the perceptual faculties infer other properties of the object, such as its shape, size, distance, and so forth, namely the common sensibles, which are sensed by other perceptive faculties, especially touch. Therefore, colour is the sole proper object of sight. Light, in turn, is a quality not inherent to air or water, and this is the reason why it can be called, improperly, the "colour of transparency", namely because it turns potential transparency (that is, darkness), and consequently the object's colours, into actuality. Moreover, Aristotle holds that light is instantaneously present in the medium once a luminous source, such as the Sun, is present. In conclusion, light cannot be an object of sight either *per se* or as a quality of the medium, otherwise the medium itself would be visible instead of transparent.

Avicenna and Averroes advanced contrasting responses to this theory, giving light a more elaborated role in the process of vision.[19] Like Alkindi, Avicenna emphasises the capacity of light to spread through transparent media. For him, radial illumination directly empowers the colour on the object's surface to be visibly radiating in the air up to the observer's eyes. Again, for him what is visible is colour, actualised by light radiation. Thus, when Avicenna asserts that "what we call light (*quod vocamus lucem*), like what the Sun and the Moon have, is that which is seen in itself (*per se ipsum*)", this "what" is not light but irradiated colour. Avicenna, in fact, clarifies that "radiance and ray are not necessary for us to see, but the other two", namely a radiating and an irradiated object.[20] Moreover, Avicenna follows Galen in locating the seat of vision in the lenses, which are in contact with the visual spirits running in the optic nerves.[21] Yet, contrarily to Galen, he thinks that the visual spirits are inside the eye, and sense the colours on the eye surface, from where they convey the visual impression through the optic nerve up to the brain. This is, basically, the physiological theory of the spirits followed by the perspectivists, and by Dante as well.[22]

Also, Averroes emphasises the role of light in colour actualisation. For him, colour and forms have only spiritual or intentional existence (*esse intentionale*) in the medium, and the same runs for light, since it is not a body but a quality ("lux non est corpus, sed est praesentia intentionis in diaphano").[23] On these premises, Averroes corrects Avicenna by asserting that colour is precisely this "luminosity mixed with transparency" (*lumen admixtum cum diaphano*). He, however, does not conceive of light as something that the eye may receive as such from the outside. This possibility is, indeed, briefly touched by him in commenting on *De anima* II (7, 419a4) and *De sensu* 2 (437b1–7) about how, in the dark, we see luminescent shells or fishes having something luminous that takes the place of colour, but this is not true colour, as it disappears in presence of light. Averroes' idea of luminescence as "visible in itself" is in line with Aristotle, who deals with the problem in the *De sensu* by distinguishing luminescence from light.

As regards the Latin commentators, Albert the Great agrees with Averroes as regards the intentional status of colour and light in the medium and with Avicenna in considering that light activates colours to visibility.[24] To my knowledge, only in the late and perhaps spurious *Summa theologiae* he mentions that light is an object of sight *per se*.[25] Thomas Aquinas, in turn, remarks with Aristotle that the proper sensible of sight is colour, but underlines that the pupil, in receiving colour, also receives light, which is the reason of the visibility of colours. In his *Commentary on the Sentences*, however, Thomas states that "what we call light is that which is visible in itself".[26] This seems to refer to the perspectivist thesis of light as a proper of sight. However, here Thomas is questioning whether light is a substantial form, and wonders: if light *were* visible, such as radiance (*fulgor*) seems to be visible, would it be a substantial form? The answer is no, and the premise implies that light is not visible as such. Yet, radiance is indeed thought of as something visible, but this is mentioned only *en passant*.

As for the Franciscans, the encyclopaedic *On the Properties of Things* by Bartholomew the Englishman, written between 1230 and 1250, does not contain the idea of light as visible – and this may be further evidence that this idea started circulating only later, in the sixties, with Bacon's and Peckham's *Perspectivae*, where this assumption is openly and frequently reiterated and specifically referred to Alhazen. Bacon, however, mentions the idea of light as proper visible also in his commentaries to Aristotle, such as in his *Liber De sensu*, where he says: "as stated in the first book of *Perspectiva*, only light and colour are sensed by sight alone, as its proper object" ("ut dicitur primo *Perspective*, solum lux et color sentiuntur a visu tanquam objectum proprium"); Peckham, in turn, repeats that "light and colour are apprehended by sight alone"; and Witelo remarks very clearly that "visibles in themselves are only light and colour, because light is seen in itself" ("per se visibilia sunt tantum duo, scilicet lux et color: quoniam lux ex se ipsa videtur").[27] The treatise *De luce* by the Franciscan Bartolomeo da Bologna, written in the '80s, names Alhazen and states twice that light is *visibilis per seipsam*.[28] Curiously, the widespread work *De oculo morali* by the French master Pierre de Limoges mentions only once that the eye "receives light from outside and reflects it".[29] All these statements are rooted in Alhazen's all-pervading tenet that light radiating from the sun or fire or from illuminated bodies is seen *per se* according to the nature of light and the very nature of sight, as, for instance, he clearly states in this passage:

> light emanates in every [possible] direction from any luminous body, however it is illuminated. Thus, when the eye faces any visible object that shines with some sort of illumination, light from that visible object will shine on the eye's surface. And it was shown that it is a property of light to affect sight, whereas it is in the nature of sight to be affected by light. It is therefore fitting that sight sense the luminosity of a visible object only through the light that shines from it.[30]

By the end of the 13th century, therefore, the idea that light is a proper object of sight is firmly present in the perspectivist literature and starts circulating, cautiously, in other sources. It is, in conclusion, a known idea but not a commonplace idea.[31]

4 A Matter of Punctuation: *Convivio* III, ix, 6 and Light as a Proper Visible

The text to which we now turn in search of Dante's theory of light in the process of vision is part of his self-exegesis of the verses of the Canzone *Amor che nella mente mi ragiona*:

> Tu sai che il cielo sempr'è lucente e chiaro
> e quanto in sé non si turba già mai;

> ma li nostri occhi per cagioni assai
> chiaman la stella talor tenebrosa.
>
> (*Conv*. III, 0, 5)
>
> [You know the sky is always bright and clear, / and of itself is never clouded. / And yet our eyes, for many reasons, / sometimes say a star is dim].

The self-exegesis of these verses (*Conv*. III, ix, 5–11) grounds the explanation of why "the truth is at times in conflict with appearance"; Dante's aim is, in fact, to explain why his canzone qualifies as "gentle" the same "donna" (personification of philosophy) he called "proud and pitiless" in a previous ballata. The present paper does not consider the widely debated figure of the Donna Gentile in Dante's poetry but is limited to examine how the poet introduces his digression on how vision takes place, in order to show why a star may appear dim though it is always and invariantly luminous.

Significantly, in *Conv.* II, iii, 6 and xiii, 27 Dante had stated that the science of *perspettiva* is grounded on experience and mathematics and deals specifically with the "vision of the stars". Now, in the digression on vision, he is exactly considering the *visibility* of a star. This enormously distant object is visible only because of its light, which *always* stands like a dot perpendicular to the observer's eye. Now, in commencing his explanation, Dante mentions Aristotle's *De anima* book 2 and *De sensu*. In this key passage, however, all current editions of the *Convivio* present a crucial error of punctuation, as I proved elsewhere and sum up hereafter.[32]

Modern editions and translations interpret this sentence as if Dante were assuming that Aristotle had posited colour and light as proper visibles. But this tenet would imply a serious misunderstanding of Aristotle's theory of sight, since, as already said, for Aristotle colour *alone* is properly visible. Now, if the right punctuation is restored, the alleged misunderstanding disappears. Dante, in fact, makes two clear and consequential statements. First, he states that *both* colour *and* light are proper visibles. Second, he adds that for Aristotle there are other things, such as figure, number, movement, which are visibles, but only improperly, since they are perceived also by other senses, especially touch. At the end of the passage, Dante repeats the first statement, to further underline its relevance: only colour and light are proper visibles since we sense them by sight alone ("solo col viso"). The passage of *Conv.* III, ix, 6, with right punctuation, runs as follows:

> Here we should know that properly speaking only color and light are visible. As Aristotle asserts in the second book of *On the Soul* and in *On Sense and Sensible*, it is true that other things are visible, but not properly speaking, because some other sense perceives them, so that they cannot properly be said to be visible, nor, properly, tangible; and such are shape, size, number, movement, and state of rest, which we call common sensibles, things that we perceive by more than one sense. But

color and light are, properly speaking, visible because we apprehend them by sight alone and by no other sense.

[Dove è da sapere che propiamente è visibile lo colore e la luce. Sì come Aristotile vuole nel secondo dell'*Anima* e nel libro *del Senso e Sensato*, ben è altra cosa visibile ma non propiamente, però che [anche] altro senso sente quella, sì che non si può dire che sia propiamente visibile né propiamente tangibile: sì come è la figura, la grandezza, lo numero, lo movimento e lo stare fermo, che sensibili [comuni] si chiamano: le quali cose con più sensi comprendiamo. Ma lo colore e la luce sono propiamente; perché solo col viso comprendiamo ciò, e non con altro senso].

Patently, this passage establishes the central issue of Dante's theory of vision and light: the assertion that light – and not colour alone – is a proper sensible of sight. The philological evidence of the new punctuation is twofold. First, almost all the manuscripts of the *Convivio*, included the three extant 14th-century copies, agree in presenting a break and/or a period before "sì come Aristotele vuole" ("as Aristotle wants"), with the "S" of "sì" in capital letters. Second, Dante often uses the syntagm "sì come" at the beginning of a sentence, starting with the very opening of the *Convivio*: "Sì come dice lo Filosofo nel principio della Prima Filosofia" (*Conv.* I). Moreover, Dante's words parallel a statement of Roger Bacon's *Perspectiva* which also refers to *De anima* book 2 and *De sensu* as the works where Aristotle examines the common sensibles: Dante lists exactly the same six common sensibles mentioned by Bacon.[33] Alessandro Parronchi noted this striking parallelism in his *Dante e la perspettiva*, but overlooked its significance, and failed to envisage the error of punctuation in Dante's passage.[34]

The statement that light and colour are visibles *per se* is, clearly, repeated several times in Bacon's *Perspectiva*. Thus, this work may well be the source of the entire opening passage of Dante's digression on vision.

Dante might have grasped the idea of light as proper visible also from Peckham's or Witelo's *Perspectivae*, which were written and circulated in Italy. Both works are structured as a sequence of brief propositions, followed by geometrical explanations. Eventually, Dante might have also accessed Alhazen's *Perspectiva*, which, again, prefaces his technical explanations with clear and reiterated statements, such as the ones reported in the section below. Alhazen's *Livre des Regardz* is mentioned in Jean de Meun's continuation of the *Roman de la* Rose, composed about 1268–1278, while the mid-14th century vernacular translation, entitled *Libro de li aspetti*, testifies that Alhazen was early read by the Italian lay audience.[35] A second-hand understanding of the idea of light as properly visible might also have reached Dante indirectly, through the sources that disseminated this perspectivist idea. It is less likely, however, that it came from the Aristotelian commentaries, where, as we have seen, it is basically absent. But how is light sensed by sight? The next section focuses on this issue and Alhazen's geometrical intromissionism.

5 Visual Certitude by Straight Line, and the Eye Pain

For Alhazen, as Mark Smith asserts,

> the science of optics depends on two core assumptions: first, that vision is due to something formal passing from the object to the eye, and second, that what passes to the eye follows imaginary straight lines to reach it. Light, according to Alhazen, is just such a thing.[36]

Light irradiation progresses along straight lines entering the eyes and producing an immediate effect on the visual organ, the pupil, which suffers great pain and temporary blindness if light is excessive. This is exactly the sign that sight is perceiving light, and, with it, colours reverberated from the object:

> sight is constituted in such a way as to suffer the effects of light and colour along a specific set of straight lines that alone intersect at its [*i.e.* of the eye] center and are perpendicular to its surface. It is granted by mathematicians that sight perceives visible objects along only those straight lines whose endpoints meet at the center of the eye.[37]

These lines are the "radial lines". Now, vision is achieved only when the eye receives light, colour, and the forms of the object along these radial lines. Hence, the shortest straight line between the object and the eye, which is perpendicular to the eye passing through the pupil, will be the path through which the strongest sensation happens, as it conveys the objectual forms, light, and colour in their best assessment. Alhazen says:

> the eye senses the light and color on the surface of a visible object by means solely of a form that extends from the surface of the visible object to the eye through a transparent medium [...], and that sight perceives only those forms [reaching the eye] along the straight lines [...] being perpendicular to all the surfaces of the tunics of the eye. This is therefore how vision takes place generally, because, in terms of naked sensation, sight perceives only the light and color that are in the visible object. The remaining characteristics of visible objects that sight perceives, e.g., shape, size, and the like, are perceived by sight not through naked sensation but through reason and defining features.[38]

The linear rectitude is, therefore, the main criterion of visual certainty, and the eye pain in receiving light – together with colour – is the specific sign of its reception according to the perpendicular line.

Dante had invoked this principle since his early *Vita Nuova* and *Rime* with reference to Beatrice's eye rays/light entering his eyes, as seen in Section 2 of this paper. The same principle is invoked in the dramatic scene of his soul blaspheming the hour when his eyes met Donna Gentile's frontal gaze, a fatal event causing the attraction of his soul, in *Conv.* II, ix, 4–5:

> The first [thing that the soul says to the poet's eyes] is that she curses the hour when this lady looked on them. Here it should be known that although many things can enter the eye at the same time, nevertheless that which enters along a straight line (*che viene per retta linea*) into the center of the pupil is the only one that is truly seen (*veramente si vede*) and which stamps itself upon the imagination. [...] for just as the one which looks receives the form in the pupil along a straight line (*riceve la forma nella pupilla per retta linea*), so along that same line its own form proceeds into the one it looks at; and many times along the extension of this line is discharged the bow of him against whom all arms are light. Therefore when I say that such a lady looked on them, it is as much as to say that her eyes and mine looked upon one another.

The dismay (*smarrimento*) of the reciprocal frontal gaze provokes in Dante a sort of feeling (*consentimento*) of effimeral love (*spiritel d'amore*), as this is the same path along which Love shoots its arrows. This happens because the straight line is that of visual certitude, through which what is seen has the strongest effect on sight.[39]

Already in the *Vita Nuova* V, 1–2, Dante had narrated that he, inadvertently, intercepted the Gentle Woman's eye intruding between "the straight line that departed from the most gentle Beatrice" ("la linea retta che movea da la gentilissima Beatrice"). Dante was, in fact, in search of Beatrice's gaze:

> It happened one day that this most gracious of women was sitting in a place where words about the Queen of Glory were being listened to, and I was positioned in such a way that I saw my beatitude. And in the middle of a direct line between her and me was seated a gracious and very attractive woman who kept looking at me wondering about my gaze, which *seemed to rest on her*. Many people were aware of her looking, and so much attention was being paid to it that, as I was leaving the place, I heard people saying, "Look at the state he is in over that woman". And hearing her name I understood they were talking about *the woman who had been situated midpoint in the straight line that proceeded from that most gracious lady, Beatrice, and reached its end in my eyes*.

In *Conv.* II, xv, 4, the principle of the certainty of the vision received through the perpendicular to the eye is again adopted to explain how the soul cannot but fall in love with the "eyes", that is, the demonstrations, of philosophy, which arrive straight into the eyes of the mind: "the eyes of this woman

[Philosophy] are her demonstrations, which, straight into the eyes of the intellect, fall in love with the soul".

We can now turn to the issue of why a star may appear "dim" if it is always "bright". If the straight line is the principle of certitude of what is seen and light is sensed per se by sight, how is it possible that the vision of a star, which is always direct and "true", maybe deceiving? After considering the distance from the object, which shrinks but does not alter the perceived object, Dante considers the medium through which the form "comes to the eye", in *Conv*. III, ix, 9:

> And so for vision to be true (that is to say, to be able to see a thing precisely as it is in itself), the medium through which the form reaches the eye must be colorless, and so too the water of the eye; otherwise the visible form would be tinged with the color of the medium as well as that of the pupil.

Thus, distance apart, only aerial turbulences and eye diseases may cause an alteration in the perception of a frontal luminous object. In the case of pure radiating light, as that of a star, there cannot be deception of sight, unless because of the impurity of the medium or the observer's sensitive organ.

6 'Seeing the Light' as a Powerful Source of Metaphors

As I have tried to illustrate, it is noteworthy that Dante embraced the perspectivist theory of light as visible in its own right. This is a notion that, while evident to us today, was pioneering and far from commonly recognised in his time. Dante, however, did not embrace this innovation in pursuit of scientific knowledge in geometric optics. Indeed, this and the related perspectivist principles of the straight line and the ocular pain are a few, simple, and non-technical principles that fit perfectly into a much higher goal, namely their being transformed into a powerful source of metaphors that permeate Dante's poetic production.[40]

In the *Vita Nuova*, the radiance emanating from Beatrice's eyes serves as a symbol of the intensity of love that profoundly affects Dante, captivating his senses and stirring his admiration. It symbolises an otherworldly and beatific love that she shares with heaven, the focal point of her gaze. Dante elevates this luminous love by portraying it as a sign of bliss, sealed by Beatrice's death. In the concluding sections of the *Vita Nuova*, and in the *Convivio*, the light radiating from the eyes of the Gentle Woman – inadvertently crossed in a face-to-face exchange of gazes – stands as a clear metaphor for the persuasive force of the philosophical demonstrations, which enamour the mind.

It would be simplistic to reduce all light metaphors presented by Dante as an expression of adhesion to perspectivism! Clearly, light has always exerted a strong, perhaps the strongest, symbolic force, especially for expressing how the divine manifests itself to the gaze of the mind. Dante extrapolates his references to light not only and not primarily from the perspectivism literature, but from the Bible, the Church Fathers, and basically from all ancient and medieval literature, where light metaphors are pervasive. Still, he considers light metaphors not only as the most efficacious way to signify God, the divine, the mind illumination, and so on but also, and above all, the most evident sign of how human beings are driven to look at him face-to-face, exactly because light is "proper" of sight.

This specific goal also emerges in light metaphors that clearly adopt the Aristotelian conception of light as an 'activator' of colours, such as in the complex double-metaphor of *Conv*. I, i, 14–15. Here, light symbolises bread, that is Dante's explanation of his love songs, which are his banquet's victuals, in turn, symbolised by colours. In this double metaphor Dante adopts the Aristotelian function of light/bread/explanation as what manifests the colours/foods/verses of his poetry. Yet, Dante's simile develops further in *Conv*. I, xiii, 12, where the bread becomes a quasi-biblical "nuova luce", an independent and luminous source of knowledge for whoever lies in shadow:

> This commentary shall be that bread made with barley by which thousands shall be satiated, and my baskets shall be full to overflowing with it. This shall be a new light, a new sun which shall rise where the old sun shall set and which shall give light to those who lie in shadows and in darkness because the old sun no longer sheds its light upon them.

In the *Commedia* the "new light" is no longer the bread/commentary, but the poem itself. Here, light metaphors and, among them, the specific image of light irradiation as an object of vision are strategically and consistently adopted to illustrate the progress of the pilgrim's ascent.

In the realm where light is absent – the dark hell – Dante recounts a solitary episode bathed in light: the vision of the castle of the great souls in the Limbo, illuminated by a huge fire. It is this fire – and not any presumed rays emitted from his eyes – that wins over the perpetual darkness (*Inf.* IV, 68–69: "ch'emisperio di tenebre vincia") and enables him to gaze straight ahead at the "philosophical family". Dante meticulously describes his staying upright, distant, and slightly elevated in relation to the philosophers (vv. 130–131: "poi ch'innalzai un poco più le ciglia"), so to receive a clear-cut image of Aristotle "surrounded by the philosophical family" ("vidi 'l maestro di color che sanno seder tra filosofica famiglia"). Dante's gaze remains detached, not in love. Despite a fleeting moment of admiration and gratitude, he moves on with his family of poets.

Among the realms of the afterlife, the Purgatory is the place where Dante watches with his human eye, still in need of regaining a perfect sight. Here, Dante's visions take on traits of realism, and even the visions of celestial creatures are depicted through the capacities of human sight. For example, the recognition of approaching figures, such as the helmsman angel (*Purg*. II, 13–42) or the seven candlesticks (*Purg*. XXIX, 44–45), happens according to the natural sequence of different perceptive stimuli: first light, then colour, that is the two proper sensibles, and finally the formal details. Moreover, in Purgatory Dante's (and Virgil's) guide is also a real object, the sun, symbolising the divine light that Dante perceives but cannot look at directly. The divine light dazzles him through the angels, who mirror God's light, just as water mirrors the sun rays. The episode of the angel of mercy in *Purg*. XV hinges on the optical law of the angular equality between the incident and reflected rays with the perpendicular to the point of reflection (vv. 16–20). This is one of the most significant 'experiences' among the many topical narrations of the pain caused by gazing at the reflected ray, as the manifestation of the irradiated divine light.[41] This extraordinary, reflected light is far more unbearable to Dante's eye (vv. 22–23: "così mi parve da luce rifratta / quivi dinanzi a me esser percosso") than the sunbeams that the poet had just shielded from his sight (the "solecchio").

Only after his cleansing in the purgatory rivers, the attainment of a renewed sight ("rinnovata vista") allows Dante to satiate his ten-year thirst (*Purg*. XXXII, 2: "la decenne sete") and behold the divine light reflected in Beatrice's eyes. Beatrice, however, does not engage in a reciprocal gaze with Dante; her eyes remain steadfast and perpetually directed towards the source of her bliss, an unwavering focus from which no blessed can or will turn away. This intellectual light is both "visible" (*Par*. XIV, 13: "la luce onde s'infiora vostra sustanza") and "true" ("luce verace", as Cacciaguida's light), in the same way as physical light is the visible and certain object of sight. Seeing and reflecting God's light is the metaphorical mark of the angelic and beatific state.

7 A Concluding Remark on the Extraordinary Grace of Seeing the Light

As Dante ascends from heaven to heaven, he gradually acquires the ability to sustain the radiance of this overriding reflected light. This divine reverberation manifests itself through various figurations, colours, images, and forms of the diverse theophanic manifestations. In the transfigured yet precise language of the *Paradiso*, the divine "living light" is the apical object for the human mind's gaze, as it blazes so bright into the eye that it cannot be sustained, (*Par*. XXIII, 31–33: "per la viva luce [of Beatrice's eyes] trasparea la lucente sustanza tanto chiara nel viso mio, che non la sostenea"). But once Dante is ready to leave the physical universe and enter the Empyrean, he must overcome the unbearable dazzle of that radiance, which awaits every soul destined for paradise: "thus did a living light shine all around me, leaving me so swathed in the veil of its effulgence, that I saw nothing else" (*Par*. XXX, 49–51). Only after this extreme sight experience, Dante's eye becomes able to gaze even the most intense light. In *Par*. XXX the verb "to see" is ubiquitous, as the radiating divine light is now the exclusive object of vision ("fassi di raggio tutta sua parvenza").

It is extremely interesting that Dante, in the Empyreon, needs a special grace, reserved to his human eyes, to see the terrestrial bodies of the blessed.[42]

Free from any medium, spatial constraints, and limits of distance – the last geometric parameter is absent in the infinity of the beatific heaven – everything becomes "truly visible" to Dante's eyes ("la vista mia nel largo e nell'ampiezza non si smarriva"). God's splendour ("isplendor di Dio") is the glory of the truthful reign ("l'alto trïunfo del regno verace"), where the Creator is directly *visible* to the creature ("che visibile face il Creatore alla creatura"). A sublime *terzina* clarifies how God manifests itself as light of an overabundant love, true good, and pure joy (*Par.* XXX, 40–42: "luce intellettüal, piena d'amore; / amor di vero ben, pien di letizia / letizia che trascende ogne dolzore"). The exposition to this light enables Dante to perceive the cosmic order, ubiquitously pervaded by God's glory. He realises that in analogy to how dense bodies constrain the reverberation of natural light, every creature, equally infused with divine light, reflects it according to its individual capacity "for the light of God so penetrates the universe, according to the fitness of its parts to take it in, that there is nothing can withstand its beam" (*Par.* XXXI, 22–24). Clearly, the perspectivist tenet of light as proper visible finds its most profound foundation here. If light were, in Aristotelian terms, merely confined to actualising the transparency of the air and manifesting colours, it would crumble Dante's highest metaphor and goal of his poetical, theological, and intellectual journey, that is, the human presumption – sustained by an overabundant grace – to gaze upon eternal light (*Par.* XXXIII, 82–83: "abbondante grazia ond'io presunsi ficcar lo viso per la luce eterna").

The ultimate fulfilment of Dante's quest is to look at the divine light in the manner of angels and blessed, who do so "according to the straight ray" ("per via di diritto raggio"). He requires the guidance of Bernard, the mystical force accompanied by the Virgin Mary's grace, which encourages and sustains the poet in "ascending the ray" ("rimontar per lo raggio"). When God stands before Dante's eyes as his frontal object of vision, the Trinitarian mystery initially manifests itself through the geometric forms of the circle, the dot, and the enigmatic, sublime apparition of the human form. But the progressive enhancement of the gaze is eventually deprived of figurations and can enter "through the ray of the exalted light that in itself is true" (*Par.* XXXIII, 52–54: "ché la mia vista, venendo sincera, e più e più intrava per lo raggio de l'alta luce che da sé è vera"). The living light endows Dante with the capacity to sustain, for an instant, the "true visible" ("luce che da sé è vera") – the resplendent light that eventually permeates his sight, senses, and mind, embracing him with the Love that moves the universe.

Notes

1 Smith, *From Sight to Light*, ix.
2 Smith, *From Sight to Light*, ix. For the summary of the theories of vision, I mainly follow Smith's volume, Lindberg, *Theories of Vision*, Federici Vescovini, *Le teorie della luce*; Tachau, *Vision and Certitude*, and Wade, "Visual Motion", in the present book.
3 Smith, "Introduction" to Alhacen, *De aspectibus*, cxv.

4 The term appears for the first time in James of Venice's translation of Aristotle's *Posterior Analytics* and was adopted by Grosseteste in his Commentary on that Aristotelian work and in the *De iride*, written about 1225–1228. See Panti, "The Oxford-Paris Connection".

5 On Alhazen's optical theory and its reception in the Latinate world see Smith, "Introduction" to Alhacen, *De aspectibus*. Hereafter, I use the usual spelling "Alhazen" instead of "Alhacen", adopted by Smith in his numerous studies on the Arab scientist and his optical science. On the early use of the term *perspectiva* see Panti, "The Oxford-Paris Connection".

6 On the presence of extramissionism among later scholastic masters see Lička, "The Visual Process".

7 For further bibliography on the early perspectivist literature and its Franciscan milieu see Raynaud, *Optics*, and Lindberg's introduction to Roger Bacon, *Perspectiva*. For Grosseteste's *De iride* see again Panti, "The Paris-Oxford Connection".

8 John Peckham's *Perspectiva communis* and *Tractatus* are both edited by Lindberg. For Witelo and the Papal milieu see Paravicini Bagliani, "Witelo".

9 For the earliest commentators' references to Dante's use of *perspectiva* in his *Commedia* see Galli, "Visual Perception", in this volume. On Dante's knowledge and use of the perspectivist literature see Parronchi, "Dante e la perspettiva". More recently, Boyde (*Dante Phylomithes*, 207) states that the *perspectiva* taught Dante "to use some of the more common words for light in their technical acceptations". See also Akbari, *Seeing through the Veil*, 141–142, on Witelo's *Perspectiva* as a source of Dante. The remarkable book by Gilson, *Medieval Optics*, scrupulously examines Dante's possible sources for optics and light, and considers the Aristotelian commentaries, especially those by Albert the Great and Thomas Aquinas, as sufficient sources since (at 55) "the optical information used by Dante was so commonplace that it is not necessary to adduce more technical and less widely diffused sources such as the *perspectivae*". See also Gilson, "Visual Theory". On Dante's early extramissionism see again Gilson, *Medieval Optics*, 39–47 and Akbari, *Seeing through the Veil*, 147–153.

10 For Alkindi see Travaglia, *Magic, Causality*, 53–66.

11 For Alfarabi's harmonisation of previous sources see Martini Bonadeo, "Ancient Philosophical Visual Theories".

12 Albert the Great, *De sensu*, I, I.5, 28.

13 All English translations from Dante's writings are taken, with some tacit changes of my own, from "The Princeton Dante Project", at https://dante.princeton.edu. The *Convivio* is translated by Lansing.

14 See, for instance, Gilson, *Medieval Optics*, 42 and Akbari, *Seeing through the Veil*, 122–123.

15 Akbari, *Seeing through the Veil*, 147–153, who also reports that Dante adopted a "fiery" extramissionism in the *Inferno*, a "watery" intromissionism in the *Purgatorio* and the perspectivist optics in *Paradiso*. I personally disagree with this interpretation, as I try to show in the present study.

16 For the *Vita Nuova* I follow the edition by Brambilla Ageno and the translation by Musa. For the *Rime*, the text is by Barbi and translation by Foster and Boyde.

17 I have added "of their instruments" for translating "de li loro instrumenti", which was missing from the English version. Properly speaking, Dante does not say that the visual spirits go out in the air, but go out of their natural abode, which is the optical nerve. They may well remain within the human body.

18 See Smith, *From Sight to Light*, 29–35. On Aristotle's theory of sight see, for instance, Sorabji, "Aristotle on Colour". Aristotle's theory is also summarised in Panti, "Il visibile", with references to the relevant passages.

19 See McGinnis, "New Light on Avicenna". See Avicenna, *Liber de anima*, 174–175 and 191: "Dicemus igitur quod lux est pars compositions huius visibilis quod

vocamus colorem". Avicenna is mentioned in *Conv.* III, xiv, 5 as having differentiated light into light proper, splendour and radiation, a difference that Dante adopts in the *Commedia.*

20 Avicenna, *Liber de anima* III, 171: "non sunt nobis necessaria radiositas vel radius sed duae primae".

21 In line with Galen, Avicenna holds that the main center of perceptive elaboration is the brain, not the hearth, as for Aristotle.

22 See Lindberg's introductions to Roger Bacon, *Perspectiva* and John Pecham, *Perspectiva communis.* For Dante, see Brigo, "A Medical Commentary", in the present volume. Yet, my analysis of Dante's references to the visual spirits contrasts Brigo's as far as the extramissionist versus intromissionist stance is involved, as examined above.

23 Averroes, *Commentarium magnum in De anima*, II, 70, 237; see also 144, 348–349, where it is repeatedly stated that sight perceives and judges colour alone. On these aspects of Averroes' theory of vision and light see Panti, "Color est lux incorporata".

24 Albert the Great, *De anima*, II, 3, 12.

25 Albert the Great, *Summa Theologiae*, I, 3, 13, 4, 46–47: "Visibile enim primum, quod est lux, sive sensibilis sive intellectualis, non fit visibile aliquo coadiuvante, sed seipso visibile est et omnis visibilis actu". On the debated authenticity of the *Summa* see the introduction by the editor, who evidences its link with Parisian theological writings of the Franciscan milieu.

26 Thomas Aquinas, *Scriptum super II Sent.*, d. 13, q. 1, a. 3: "illud, quod nos appellamus lucem, est illud quod per se videtur". See also Gilson, *Medieval Optics*, 60–61.

27 Roger Bacon, *Liber De sensu*, 25; Witelo, *Perspectiva* II, Petitio 2; John Pecham, *Tractatus*; Pecham, *Perspectiva communis*, prop. 58.

28 Bartolomeo da Bologna, *De luce*, 249 and 317.

29 On the use of geometrical optics in predication see Ledda, "Filosofia e ottica"; Galli, "The *perspectiva*" and Galli, *Il* De luce *di Bartolomeo.*

30 Alhacen, *De aspectibus*, I.6.1, 22; translation at 355.

31 Gilson, *Medieval Optics*, 61, says that "Dante's view that light is visible in its own right should thus be seen as part of a general debt to widely accepted thirteenth-century analyses of the nature of light". As I tried to show here, this tenet is questionable.

32 Panti, "Il visibile"; Panti, "Di là dal modo".

33 Roger Bacon, *Perspectiva* I, 1.3, 10: "Et sunt sensibilia communia de quorum aliquibus exemplificat Aristoteles secundo *De anima* et in principio *De sensu et sensato*, ut de magnitudine et figura, motu, quiete, et numero. [...] quia communiter ab omnibus sensibus particularibus vel a pluribus discernuntur – et maxime a visu et tactu". See Panti, "Il visibile", for a wider exam of this source.

34 Parronchi, "Dante e la perspettiva", 33.

35 For Jean de Meun see Akbari, *Seeing through the Veil*, 88–96, which states that Jean's reference to Alhazen comes from Bacon's *Perspectiva.* The *De li aspecti* was discovered by Narducci, "Nota intorno a una traduzione", and later studied for its being a relevant source of Lorenzo Ghiberti's *Third Commentary.* See, *inter alia*, Federici Vescovini, *Le teorie della luce*, and Raynaud, *Optics*, 18–20. Needless to say, the topic of light is dominant in 13th century poetry. For instance, the poets of the "Sicilian school" (about 1230–1250) ubiquitously refer to the lady's eye spirits and light, as recently examined also in Tarud Bettini, *Luce amore visione.* This, however, does not mean that they adhered to perspectivism or were influenced by Alhazen. The issue cannot be addressed here for reasons of space and limits of knowledge from my part, but an outlook on Giacomo da Lentini's references to the lady's eye light let envisage that he basically follows the Aristotelian theory

of vision, based on the reception of forms and the prevailing role of the hearth in image elaboration.

36 Smith, *From Sight to Light*, 185. Seer also the passage quoted above at note 27.

37 Alhacen, *De aspectibus* I, 6.43–44, 369.

38 Alhacen, *De aspectibus* I, 6.60–61, 374.

39 Differently from the tenet of light as proper visible, which is absent from the Aristotelian commentaries mentioned above, the principle of perpendicularity is openly recalled also in these works. For instance, Albert the Great, *De sensu et sensato* (I, I.14, 52) refers to it as a principle exposed in the section "on direct sight in the books of perspective" (*de visu in perspectivis*). This idea was also invoked in the Euclidean, Ptolemaic and Kindian geometrical optics. Indeed, it was more common than the 'extravagant' idea of light as proper visible. Dante may well have known the principle of perpendicularity through Albert, but surely did not derive from the latter its connection with the tenet of light as object of sight.

40 For the use of metaphors in Dante and related bibliography see, for instance, the recent volume by Tomazzoli, *Metafore e linguaggio*.

41 For a full exam of the episode see Gilson, "Purgatorio XV". See also Panti, "Di là dal modo", for the use of the law of reflection in this episode.

42 For the complex and debated issue of the Empyreon in Dante see Nardi, "La dottrina dell'Empireo", Pegoretti, "L'Empireo in Dante", also for further bibliographic references. For Dante's spatial displacement see Brigandì, "Il cambio di sistema".

Bibliography

Primary Sources

Albert the Great. *De anima*. Edited by Clemens Stroick. Münster: Aschendorff, 1968.

Albert the Great. *De sensu et sensato*. Edited by Silvia Donati. Münster: Aschendorff, 2017.

Albert the Great. *Summa theologiae sive de mirabili scientia Dei*. Edited by Dionysius Siedler. Münster: Aschendorff, 1978.

Alhacen. *De aspectibus*. Edited by A. Mark Smith. *Alhacen's Theory of Visual Perception: A Critical Edition, with English Translation and Commentary, of the First Three Books of Alhacen's De aspectibus, the Medieval Latin Version of Ibn al-Haytham's Kitab al-Manazir*. Philadelphia, PA: American Philosophical Society, 2001.

Averroes. *Commentarium Magnum in Aristotelis De anima Libros*. Edited by F. Stuart Crawford. Cambridge, MA: The Mediaeval Academy of America, 1953.

Avicenna. *Liber de anima seu sextus de naturalibus*. Edited by Simone Van Riet. Louvain and Leiden: Peeters-Brill, 1972.

Bartolomeo da Bologna. *De luce*. Edited by Francesca Galli. *Il* De luce *di Bartolomeo da Bologna. Studio e edizione,* 249–343. Florence: SISMEL-Edizioni del Galluzzo, 2021.

Dante Alighieri. *Convivio*. Translated by Richard H. Lansing. *Dante's "Il Convivio"*. New York: Garland, 1990.

Dante Alighieri. *Convivio*. Edited by Franca Brambilla Ageno. Florence: Le Lettere, 1995.

Dante Alighieri. *Rime*. Edited by Michele Barbi. Florence: Società Dantesca Italiana, 1960.

Dante Alighieri. *Rime*. Translated by K. Foster and P. Boyde. *Dante's Lyric Poetry*. Vol. 1. Oxford: Clarendon, 1967.

Dante Alighieri. *Vita Nuova*. Edited by Michele Barbi. Florence: Società Dantesca Italiana, 1960.

Dante Alighieri. *Vita Nuova*. Translated by Mark L. Musa. Bloomington: Indiana University Press, 1965.

John Pecham. *Perspectiva communis*. Edited by David C. Lindberg. *John Pecham and The Science of Optics, Perspectiva communis*, 59–239. Madison, Milwaukee, WI, and London: The University of Wisconsin Press, 1970.

John Pecham. *Tractatus de perspectiva*. Edited by David C. Lindberg. St. Bonaventure, New York: Franciscan Institute Publications, 1972.

Peter of Limoges. *The Moral Treatise on the Eye*, Edited by Richard Newhauser. Toronto: Pontifical Institute of Mediaeval Studies, 2012.

Roger Bacon. *Liber De sensu et sensato*. Edited by Richard Steele. Oxford: Oxford University Press, 1937.

Roger Bacon. *Perspectiva*. Edited by David C. Lindberg. *Roger Bacon and the Origins of Perspectiva in the Middle Ages. A Critical Edition and English Translation of Bacon's 'Perspectiva' with Introduction and Notes*, 1–339. Oxford: Clarendon Press, 1996.

Thomas Aquinas. *Scriptum super II Librum Sententiarum*. Edited by Pierre Mandonnet, vol. 2. Paris: P. Lethielleux, 1929.

Secondary Sources

Akbari, Suzanne C. *Seeing through the Veil. Optical Theory and Medieval Allegory*. Toronto: University of Toronto Press, 2004.

Brigandì, Ottavio. *Il cambio di sistema di riferimento quale aspetto innovativo della riflessione gnoseologica di Dante*. In *Letteratura e Scienze*, edited by Alberto Casadei, Francesca Fedi, Annalisa Nacinovich, and Andrea Torre, 1–4. Rome: Associazione degli Italianisti, 2021.

Boyde, Patrick. *Dante Philomythes and Philosopher: Man in the Cosmos*. Cambridge: Cambridge University Press, 1981.

Federici Vescovini, Graziella. *Le teorie della luce e della visione ottica dal IX al XV secolo. Studi sulla prospettiva medievale e altri saggi*. Perugia: Morlacchi, 2003.

Galli, Francesca. *Il* De luce *di Bartolomeo da Bologna*. Studio e edizione. Florence: SISMEL-Edizioni del Galluzzo, 2021.

Galli, Francesca. "The *perspectiva ad usum praedicatorum* in Late 13th-century Florence. Some Observations Based on an Examination of Codex Plut., 17 sin. 8 From the Laurentian Library". *Micrologus* 29 (2021): 181–202.

Gilson, Simon A. *Medieval Optics and Theories of Light in the Works of Dante*. Lewiston, Queenston and Lampeter: The Edwin Mellen Press, 2000.

Gilson, Simon A. "Purgatorio XV". In *Lectura Dantis VII,* edited by Emilio Pasquini and Carlo Galli, 101–115. Bologna: BUP, 2017.

Gilson, Simon A. "Visual Theory". In *The Oxford Handbook of Dante*, edited by Manuele Gragnolati, Elena Lombardi, and Francesca Southerden, 242–256. Oxford: OUP, 2021.

Ledda, Giuseppe. "Filosofia e ottica nella predicazione medievale". In *Letteratura in forma di sermone. I rapporti tra predicazione e letteratura nei secoli XIII-XVI. Atti del Seminario di studi (Bologna 15–17 novembre 2001)*, edited by Ginetta Auzzas, Giovanni Baffetti, and Carlo Delcorno, 53–78. Florence: Olschki, 2003.

Lička, Lukáš. "The Visual Process: Immediate or Successive? Approaches to the Extramission Postulate in 13th Century Theories of Vision". In *Medieval Perceptual Puzzles: Theories of Sense-Perception in the 13th and 14th Centuries*, edited by Elena Băltuță, 73–110. Boston and Leiden: Brill, 2020.

Lindberg, David C. *Theories of Vision from Alkindi to Kepler*. Chicago, IL and London: Clarendon Press, 1976.

Martini Bonadeo, Cecilia. "Ancient Philosophical Visual Theories and Their Use by al-Fārābī". *Ishraq* 10 (2022): 13–38.

McGinnis, Jon. "New Light on Avicenna: Optics and Its Role in Avicennan Theories of Vision, Cognition and Emanation". In *Philosophical Psychology in Arabic Thought and the Latin Aristotelianism of the 13th Century*, edited by Luis Xavier Lopez-Farjeat and Jörg Tellkamp, 41–57. Paris: Vrin, 2013.

Nardi, Bruno. "La dottrina dell'Empireo nella sua genesi storica e nel pensiero dantesco". In Bruno Nardi, *Saggi di filosofia dantesca*, 167–214. Florence: La Nuova Italia, 1967.

Narducci, Enrico. "Nota intorno a una traduzione italiana fatta nel secolo decimoquarto del trattato d'ottica d'Alhazen". *Bollettino di bibliografia e di storia delle scienze matematiche e fisiche* 4 (1871): 1–40.

Panti, Cecilia. "The Oxford-Paris Connection of Optics and the Theory of Rainbow: Grosseteste's *De iride*, pseudo-Oresme's *Inter omnes impressiones* and Bacon's *Perspectiva* in Paris, BnF, lat. 7434". In *Le Moyen Âge et les sciences,* edited by Danielle Jacquart and Agostino Paravicini Bagliani, 251–280. Florence: SISMEL-Edizioni del Galluzzo, 2021.

Panti, Cecilia. "Il visibile 'per retta linea': Dante, la *perspettiva* e una proposta di emendamento in *Convivio* III ix 6". In *Dante e il mondo. Tra realtà e poesia tra storia e letteratura. Atti del LVIII Convegno storico internazionale. Todi, 10–12 ottobre 2021*, 143–180. Spoleto: Centro Italiano di Studi sull'Alto Medioevo, 2022.

Panti, Cecilia. "'di là dal modo che 'n terra si vede'. Dante, la luce e l'esperienza del vedere, aldiquà e aldilà del cielo". In *L'esperienza di Dante. Atti del Convegno internazionale di Studi Bologna-Ravenna, 12–13 novembre 2021*, edited by Giuseppe Ledda, 37–66. Ravenna: Centro Dantesco dei Frati Minori Conventuali, 2023.

Panti, Cecilia. "'Color est lux incorporata perspicuo'. Robert Grosseteste's Definition of Colour and Its (Franciscan) Legacy". In *Colour Theories from Democritus to Descartes*, edited by Veronique Decaix and Katerina Ierodiakonou. Abingdon: Routledge, 2025, 169–198.

Paravicini Bagliani, Agostino. "Witelo et la science optique à la cour pontificale de Viterbe (1277)". *Mélanges de l'Ecole française de Rome* 87 (1975): 425–453.

Parronchi, Alessandro. "Dante e la perspettiva". *Studi danteschi* 36 (1959): 5–103.

Pegoretti, Anna. "L'Empireo in Dante e la 'divina scienza' del Convivio". *Filologie medievali e moderne* 18 (2018): 161–188.

Raynaud, Dominique. *Optics and the Rise of Perspective: A Study in Network Knowledge Diffusion.* Oxford: The Bardwell Press, 2014.

Smith, A. Mark. *From Sight to Light. The Passage from Ancient to Modern Optics.* Chicago, IL and London: The University of Chicago Press, 2015.

Sorabji, Richard. "Aristotle on Colour, Light and Imperceptibles". *Bulletin of the Institute of Classical Studies* 47 (2004): 129–140.

Tachau, Katherine. *Vision and Certitude in the Age of Ockham: Optics, Epistemology, and the Foundations of Semantics, 1250–1345*, Leiden – New York – Copenhagen – Cologne: Brill, 1988.

Tarud Bettini, Simone. *Luce, amore, visione. L'ottica nella lirica italiana del Duecento*. Rome: Aracne, 2013.

Tomazzoli, Gaia. *Metafore e linguaggio figurato nel Medioevo e nell'opera di Dante.* Venice: Ca'Foscari, 2023.

7 Visual Perception in Dante's *Commedia* According to the Early Commentaries (1320–1400)

Francesca Galli[1]

As Patrick Boyde stated in the preface of Simon Gilson's monograph *Medieval Optics and Theories of Light in the Works of Dante*, "the words 'vedere' and 'luce' (and their cognates) are among the most frequent and distinctive items" in the poet's imagery and writings.[2] Furthermore, in Dante's time notions and theories about the functioning of sight and the behaviour of light rays were not only of interest to physicists and anatomists, but occupied a prominent place in the broader culture: they were also discussed by theologians, philosophers, and artists.[3] It is, therefore, unsurprising that much has been written about Dante's knowledge of and ideas on optics (understood as the science of light and vision), about his sources and reference models, and about the wonderful images and metaphors found in the *Commedia*.[4] On the other hand, with the exception of a recent contribution by Gilson, there are, to my knowledge, no systematic insights into the way optical contents are approached and employed by Dante's early commentators.[5] While rather neglected, this is not an irrelevant point. Indeed, it tells us something about the domain of expertise of medieval readers and about the reception and interpretation of certain images and theories in years close to the composition of the poem.[6] The present contribution addresses this research question, already sketched out in a previous article that, however, focused on the presence and use of analogies and concepts related to optics in preaching, particularly in Tuscany around the year 1300.[7] My aim is to conduct a preliminary survey and lay the foundations for subsequent in-depth investigations.

1 The *Perspectiva* in the Early Commentaries

Since the range of topics is vast, it is essential to circumscribe the field of inquiry. First, I chose to define a time frame and thus to consider those commentaries that date from the period between Dante's death and the beginning of the following century (1400).[8] The definition of the *terminus ante quem* is certainly arbitrary, but for the moment I have deemed it preferable to stop at the threshold of the century in which pictorial perspective was 'invented' and a significant paradigm shift took place. I then checked whether any author within this chronological time span explicitly mentions the science of

DOI: 10.4324/9781003412601-8

light and vision (*perspectiva*) and in what terms. Research carried out on the commentaries included in the DDP database and subsequent spot checks on the recent national editions, published by Salerno, and on texts not included in the database (e.g. Andrea Lancia's commentary) have revealed that at least six of Dante's first readers – namely Iacomo della Lana, the author of the *Ottimo Commento*, the so-called Amico dell'Ottimo, Benvenuto da Imola, Francesco da Buti, and the commentator known as the Anonimo Fiorentino – refer to *perspectiva*.[9] Table 7.1 provides an overview of the occurrences of the term in the above commentaries.[10]

Of course, the references we find in a commentary may derive from an earlier text, e.g. the *Ottimo* and the Anonimo Fiorentino often depend on Lana. Nevertheless, further scrutiny of key words and particularly significant passages (e.g. some similes dealing with light and vision) suggests that all these commentators had a non-superficial interest in optics and related topics. In the selected excerpts, the terms "perspettiva"/"prospettiva" may designate the science itself or a specific treatise, in which case the term is capitalised in modern editions – incidentally, the most popular optics handbook in the 14th century was Peckham's *Perspectiva communis*. It should also be borne in mind that this science is applied in different contexts and to a variety of subjects. In medieval culture, as already remarked, it was a multi-faceted field of knowledge, at the crossroads between various disciplines, including geometry, physics, medicine, and astronomy. In the *Convivio* (II, xiii, 25) Dante says that geometry has optics as its handmaiden ("la sua ancella, che si chiama Perspectiva"), and Benvenuto da Imola specifies that it concerns both geometry and philosophy.[11]

Similarly and consequently, in Dante's verses and related commentaries optical science is called into question with regard to multiple issues: the (mal–)functioning of sight (*Inf.* X), the properties of mirrors (*Inf.* XXIII and *Purg.* XXV), and the optical effects and illusions whereby something appears to be bigger or smaller than it actually is or is mistaken for something else (*Inf.* XXXI: towers/giants; *Purg.* XXVII: the dimension of stars; *Purg.* XXIX: trees/candelabra). In *Purg.* XV, 21, the poet himself mentions the "arte" (i.e. the discipline), when in a rather technical manner he dwells on the movements of light rays and the equality of the angles of incidence and reflection. Furthermore, in *Purg.* XXXIII and *Par.* II optical knowledge is evoked with respect to astronomical matters. At the end of the second *cantica* and with regard to how the sun appears to move more slowly when it is close to the zenith, Iacomo della Lana (as well as the Anonimo Fiorentino) mentions a *Perspetiva*, along with the *Canons to the Toledan Tables*.[12] Finally, the second canto of *Paradiso*, where the author lingers on moon spots and describes the three-mirror experiment, is very rich in optical content, as is also evidenced by the splendid miniature by Giovanni di Paolo (ca. 1450) in ms London, British Library, Yates Thompson 36, fol. 132.

Given the main purpose of this volume, I will focus on those passages where image transmission and visual perception are discussed and leave for

Table 7.1 Occurrences of the term *perspectiva* (in Latin and the vernacular) in Dante Early Commentators

Iacomo della Lana (1323–1328)	*Ottimo Commento (c1330–1334)*	*Amico dell'Ottimo (c1341–1343)*	*Benvenuto da Imola (c1379–1383)*	*Francesco da Buti (c1390–1396)*	*Anonimo Fiorentino (14th ex.–15th in.)*
Inf. X, 100–105: Prospectiva / Perspectiva	*Inf*. X, 100–105: Prospectiva	*Inf*. X, 100–105: Perspectiva	*Inf*. X, 100–105: perspectiva	*Inf*. X, 61–62: Prospettiva	
Inf. XXIII, 25–33: Perspettiva / Perspectiva			*Inf*. XXIII, 25–27: Perspectiva		*Inf*. XXIII, 25–27: Arte prospettiva
				Inf. XXXI, 19–27: Prospettiva	
Purg. XV, 1–6 and 16–23: Prospetiva / perspettiva			*Purg*. XV, 21–24: Prospectiva	*Purg*. XV, 16–33: Perspettiva	*Purg*. XV, 1–6: Prospettiva
				Purg. XXV, 16–30: Perspettiva	
Purg. XXVII, 89–90: Perspetiva / perspettiva					*Purg*. XXVII, 89–90: Prospettiva
Purg. XXIX, 46–50: Prospetiva / perspettiva					*Purg*. XXIX, 46–48: Prospettiva
Purg. XXXIII, 103–105: Perspetiva / Perspettiva		*Purg*. XXXIII, Perspectiva			*Purg*. XXXIII, 103–105: Prospettiva
				Par. II, 112–126 and 139–148 Prospettiva	

future studies the commentators' arguments about light propagation, the luminosity of celestial bodies, and mirrors. The definition of the object of research will lead me to turn my attention mainly, albeit not exclusively, to verses and episodes set in hell (particularly from *Inf.* IX, X, XV, and XXXI). This is the realm where the eye is not dazzled, but rather tries to grasp something in the darkness. It is a place of mist and gloom ("nebioso e scuro"), unlike paradise, which is all light ("per contrario dovei è tutta la luxe, ch'è 'l paradiso"), as Iacomo della Lana points out.[13] As will be seen, various cues already included in the poem and others inserted quite independently in the commentaries will lead Dante's readers to deal with the anatomy of the eye and the physiological causes that alter visual perception, with the theories of vision (extramission *vs* intromission), and with geometric optics and the spatial conditions required for the correct reception of images (clear air, an adequate distance).

2 The Faulty of Eyesight and Its Causes

The first passage which practically compels commentators to engage with the science of vision is in *Inferno* X[14]:

> 'Noi veggiam, come quei c'ha mala luce,
> le cose', disse, 'che ne son lontano;
> cotanto ancor ne splende il sommo duce.
>
> Quando s'appressano o son, tutto è vano
> nostro intelletto; e s'altri non ci apporta,
> nulla sapem di vostro stato umano'.
> (*Inf.* X, 100–105)

> ['We see, like those with faulty vision, / things at a distance', he replied. 'That much, / for us, the mighty Ruler's light still shines. / When things draw near or happen now, / our minds are useless. Without the words of others / we can know nothing of your human state'].

Shortly before (vv. 67–72), Dante, puzzled by the question about Cavalcante's son (i.e. whether Guido was still alive or not), had hesitated to answer it. Cavalcante, distraught, had fallen down supine and then disappeared. Dante now voices his doubt as to what the damned see – that is, what they know. He guesses they have some awareness of the future but wonders what they glimpse of the present and asks Farinata to shed some light on this conundrum: "pray untie for me this knot" (*Inf.* X, 95: "solvetemi quel nodo"). Farinata replies by formulating the famous comparison with those who have "mala luce". He points out that the damned are precluded from learning about events close in time, unless they receive news from others and conclude that they will no longer know anything with the arrival of the last day: "all our knowledge / will perish at the very moment / the portals of the future close" (vv. 106–108: "tutta morta fia nostra conoscenza da quel punto che del futuro fia chiusa la porta").

The parallel with those who have poor eyesight is variously explained by ancient and contemporary commentators. Among the latter, most (e.g. Anna Maria Chiavacci Leonardi) explain that the damned are similar to presbyopes, in the sense that they see well from afar and poorly up close.[15] Some, however, e.g. Singleton, propose a slightly different interpretation and claim that the meaning of Farinata's words is that the inhabitants of hell see "dimly" at a distance (and nothing of things near them): "their light, which is said to come from God, is 'poor' only in the sense of being insufficient".[16] Early commentators are essentially divided between two explanations for Farinata and his companions' defective eyesight, one referring to the principles of geometrical optics and the other focusing on physiological causes. Iacomo della Lana is a spokesman for the former view. He makes it clear, also by referring to the commentaries on Aristotle's *Physics*, that Farinata uses the example of the sense of sight to speak of intellectual knowledge. He then quotes Aristotle's *Metaphysics* ("sensibile super sensum conrumpit sensum") and considers the question in spatial terms, emphasising that there must be an adequate distance ("proportionevele distancia") between the eye and the object in order for the latter to be seen. To prove his point, he mentions a second *auctoritas*, Euclid's *Optics* ("come ne demostra Euclide in la *Prospectiva*") and explains that the visual process takes place according to a triangular pattern ("ell'è neccesse che ongni cosa che sende se veça per triangollo"): it can be conceived of (and represented) as a cone or a triangle with the eye as its apex. Unlike the future, which is yet to come, the present time for the damned is like an object placed directly on the eye ("è come fusse in su l'ochio"), so close that the visual triangle cannot be formed ("quando la cosa è sovra lo ochio, ella non pò costituire triangolo"); hence, knowledge is precluded to them ("non se vede").[17] The *Ottimo* and thus also the Amico dell'Ottimo take up Lana's exposition almost literally, although the Latin quotations from Aristotelian works are in the vernacular here. This is the only passage in these two commentaries in which the term *Prospectiva/Perspectiva* is explicitly used.[18]

Many readers, on the other hand, overlook the geometric-spatial aspects of visual perception and dwell instead on the gradual loss of the faculties of the human eye with age. Pietro Alighieri compares the damned to ageing men ("senescentes homines"), whose visual power ("visus, idest virtus visiva") thickens and weakens ("ingrosatur et incipit debilitari"), becoming less effective.[19] Roughly the same remark can be found in the *Chiose cagliaritane*: "those with bad eyesight are obviously old people, who see better from a distance than up close".[20] Immediately afterwards the anonymous commentator – like Dante in *Conv.* III, ix, 14 – describes the gesture of an old man who cannot see what is near him and moves the object away from himself in order to see it better ("quando l'omo vecchio vole vedere una cosa sì la se delunga dal viso et alora la vede et apresandolase nol la vede"). Boccaccio also offers a physiological explanation. He speaks of a defect due to ageing ("questo vizio avvenire agli uomini quando vengono invecchiando"), when bodily fluids coming from the brain ("omori li quali vengon dal cerebro") block the visual power and prevent one from seeing things nearby ("l'occupano intorno alla vista delle cose propinque"). Only when the *virtus visiva* extends farther ("la virtù visiva si stende più avanti"), the dimness ("adombrazion dell'omore") is lessened and the quality of perception improves ("men mal vede e con più sincerità").[21]

A similar reference to the visual rays gradually brightening as the distance from the eye increases occurs in Benvenuto's commentary. The master from Imola clarifies that the damned see the future as those with weak sight see things ("damnati vident futura eo modo, quo ille qui habet debilem visum videt rem aliquam"). Then, explicitly referring to the teachings of optical science ("cuius ratio assignatur in perspectiva"), he says that eyesight can be defective in two ways, as there are those who see well from afar but poorly at close range and those who are in the opposite condition ("aliqui vident rem melius de longinquo, alii de propinquo"). The latter have limpid but limited vision ("habent visum clarum, sed non multum"), while the visual rays of the former become clear only after travelling a long way ("radii visuales egrediendo serenantur").[22] The damned belong to the former category ("autor hic loquitur de illis qui vident rem a longe, sed non de prope"), and can therefore know the future but not the present.[23]

Both the master of Imola and Boccaccio would later return to the subject of weakened and blurred vision in relation to *Inf.* XV, 20–21, when commenting on those souls who "peered with knitted brows" ("aguzzavan le ciglia") at Dante and Virgil like an elderly tailor threading a needle. Boccaccio explains that when eyesight does not work properly ("difetto degli spiriti visivi"), for example owing to a certain "thickness" ("grosseza"), the act of peering with knitted brows ("aguzar le ciglia") condenses visual power into a narrower space ("ristrignamo in minor luogo"), thereby making it sharper and more effective ("più acuta e più forte al suo uficio"). An identical explanation of the tailor's gesture is given by the Amico dell'Ottimo ("strigne<ndo> le ciglia, per la vista degl'occhi più ragunati insieme, più adoperi").[24] Benvenuto expounds the same concepts and recalls that the issue in question is addressed in the (pseudo)Aristotelian *Problemata* ("Aristotiles libro Problematum quaerit rationem huius"), taking the example of archers 'squinting' ("balistarii hoc faciunt") to direct arrows at their target.[25]

Finally, if we return to Farinata's words in *Inf.* X, we can observe that the Anonimo Fiorentino, who often refers to Lana, opts instead for some anatomical-physiological clarifications in this passage. A few details are added to what had already been written by his predecessors. The commentator notes that there are two natural reasons ("per defetto naturale") as to why eyesight may prove faulty, especially as humans age ("attempati"). The visual spirit may be too thick ("ingrossato") or, on the contrary, too thin ("troppo assottigliato"). In the latter case, it is too weak ("debole et fragile") to reach objects in the distance, whereas it allows the perception of near things. Conversely, in the former case, it benefits from moving away from the eye. Indeed, as the *virtus visiva* stretches ("dilungandosi"), it becomes sharper, thinner, and more united ("s'appunta et assottigliasi et è più unita"); in addition, the amount of air, and thus of light, between the sensory organ and the object increases ("v'ha più aere et per seguente più lume"). As a consequence, the Anonimo concludes, some people tend to hold reading material farther away to make the letters clearer.[26]

3 Echoes of the Debate about the Visual Process (Intromission or Extramission?)

Unlike the other commentators we have considered, Francesco da Buti quickly moves beyond the comparison developed by Farinata, and instead finds a cue to delve into optical matters in the dialogue between Dante and Cavalcante (*Inf.* X, 52–72).[27] Indeed, Guido's father is troubled by the past tense of the verb 'to have' used by Dante: "What? / Did you say 'he held'?" (vv. 67–68: "Come? / Dicesti 'elli ebbe'?"). He wonders whether his son is still alive: "Lives he not still?" ("Non viv'elli ancora?"). The question is rephrased a second time in v. 69: "Does not the sweet light strike upon his eyes?" ("non fiere li occhi suoi lo dolce lume?"). Alongside this first version (and interpretation) of the text, the commentator presents an alternative reading, based on the *varia lectio*: "Altro testo dice: *Non fieron li occhi etc*".[28] The eyes thus become the subject, and the plural verb ("fieron") would indicate an action of the visual organ ("li occhi veggono mettendo fuori li raggi visuali"). In this case – and the author seems to have a specific treatise in mind ("come dice nella Prospettiva") – the visual rays, actively emitted by the eye, reach the object and are then reflected through the mediation of light ("percossi nella cosa veduta, si riflettono alli occhi mediante la luce"), bringing the image back to the visual organ ("rapportano all'occhio").[29]

New thoughts on this point are offered by Francesco when commenting on *Purg.* XXIX.[30] The reason why candelabra appear to be golden trees is distance ("distanzia"), for if the space between the object and the eye exceeds the visual organ's ability to perceive ("per lunghessa del mezzo che eccede tanto la virtù visiva"), the sense is deceived ("'l vedere s'inganna"). Going into more detail, the commentator explains that the rays emitted by the eye shrink back before they have reached the object ("ch'e' raggi visuali si ristringingeno innanti che vegnino a l'obietto"). This clarification implies a further elucidation of the mechanism of visual perception. Francesco states that the visual rays, which are responsible for conveying images to the organ of perception ("avendo virtù di rappresentare all'occhio la cosa veduta"), are emitted by the eye ("l'occhio mette fuora raggi") and stretch out ("si dilatano") to grasp the object in its entirety, albeit within the limits of the subject's sensory powers ("tanto che vegnano al mezzo correspondente a la virtù visuale"). In this way, the eye, which is itself small ("picculino"), can see an object of considerable size, such as a tower, from top to bottom ("apprende dil sommo a la fine"). However, if the distance exceeds the capacity of vision, the visual cone, which has the eye as its apex, shrinks back ("incominciano a chiudere", "ritorna in acuto") before perception is complete and things grow smaller ("le cose che si veggiono incominciano ad apparire minori più l'una che l'altra") until they disappear ("e così non vede poi più nulla"). Without any references to geometry and Euclid, the phenomenon is graphically represented as a kind of double triangle or rhombus ("sì come appare in questa figura < >").[31]

In the continuation of his argument Francesco notes that, once Dante gets closer to the candelabra, the deception vanishes. These verses lead the commentator to consider a different theory of vision. Having reduced the distance, the object to be perceived – "commune" because it initially concerns all the senses ("la cosa da esser appresa per li sentimenti è comune a tutti e cinque li sentimenti") – does not fail in its operation and can thus be grasped properly.[32] Here it is the object, and not the eye, that plays an active role and imprints its own image ("ombra" and "imagine"), which is then received by the sensory organ ("'l vedere adoperi, ricevendo impressione da l'obietto, e non operi mettendo fuora li suoi raggi"). Francesco reports this opinion without taking any stance on it ("e questo dice secondo l'opinione di coloro che tegnano che 'l vedere etc.").

The master from Pisa is not the only reader of Dante grappling with the different attitudes towards visual perception developed in the Middle Ages, that is 'intromissionist' and 'extramissionist' theories.[33] While commenting on the term "postille" in *Par.* III, 13–15, that is reflections in a transparent body such as water or glass ("quella imagine nostra, che ci si rapresenta in acqua o in ispecchio, o altro corpo trapassante"), the *Ottimo* points out that there are two models of how vision works. The question of what moves towards what – whether the image from the object ("Della cosa che noi vedemo, cioè la sua forma viene agli occhi nostri") or rays from the eye ("o li raggi visivi vanno alla cosa veduta") – remains open, since the author claims to have dealt with it elsewhere ("è trattato altrove"), although I have been unable to identify any other passages in which the topic is further explored.[34] Nevertheless, hints about the theories of vision are made here and there. For example, in the extensive introduction to *Par.* XXXIII, the commentator, repeating Iacomo della Lana's words almost verbatim, says that the perceiver gets to know the object by the image that reaches the pupil ("quando noi veggiamo una cosa, la spezia visibile di quella cosa ne viene alla pupilla e per quella la conosciamo").[35] Like the *Ottimo*, the Anonimo Fiorentino also explicitly refers to the contemporary debate about how visual perception works. In commenting on *Inf.* VIII, 3, he observes that there are conflicting opinions among natural scientists ("Egli è gran quistione fra' fisici") as to whether sight goes towards the thing or the other way around ("se 'l nostro vedere va alla cosa, o se la cosa veduta viene agli occhi"). Yet, despite multiple opinions ("et come che molte oppenioni ne sieno"), there seems to be common agreement about the image being transmitted to the eye ("comunemente s'accordono che l'ojetto viene agli occhi").[36]

4 The Anatomy of the Eye: Speaking of "nerbo" and "gonna"

After mentioning the dispute between intromission and extramission theories, the Anonimo Fiorentino introduces a lengthy anatomical description of the visual system.[37] He says that the eye is a spherical, watery body ("corpo sperico et acquoso") with membranes or sheaths ("panniculo") that contain the fluids or humours and prevent them from falling out and dispersing ("questo licore caderebbe et disfarebbesi"). He enumerates various parts of the visual organ, providing summary information for each of them. Three membranes separate the cornea ("cornea"), the crystalline lens ("parte [...] cristallina"), and the vitreous humour ("parte [...] vitrea"). The cornea is the "white outer [part] of the eye" ("bianco di fuori dell'occhio"); it is named after the horn ("corno"), which is a bone ("osso") and is also used as a writing implement.[38] Besides these parts ("fuori di queste parti"), there is the pupil, also called "Luce". This is the black part of the eye ("nero dell'occhio") where images multiply ("ove si multiplicano le spezie visive"). Below the pupil ("sotto questa Luce") is the uvea ("Uva"), which is shaped like a grape ("come un acino d'uva") and has a hole that is adjacent to the pupil ("quello foro viene al lato alla Luce").[39] The commentator also distinguishes between those nerves responsible for sensation ("dà il sentimento della cosa") and those responsible for motion ("dà il movimento alla virtù visiva"): the former originate from the front of the brain, the latter from the back ("nucrea"). Finally, he deals with the optic nerves, which are part of a single nerve that comes from the brain and then splits into two ("procede dal cerabro uno nerbo [...], si divide in due"). As the right nerve extends to the left eye ("'l nerbo destro va all'occhio sinistro") and the left one to the right, these nerves cross ("incrocicchiasi") in the shape of an X before branching out and reaching the aperture of the uvea ("vengono infino a quel forato dell'Uva"). The optic chiasm is the site where judgments are made ("giudica") about the thing seen, for example by assessing whether it is "tall or short" ("se la cosa è alta o bassa") or "an ox or a horse" ("bue o cavallo").

The Anonimo then refers back to this thorough exposition a little later, when commenting on Virgil's words in *Inf.* IX, 73–74.[40] The Latin poet invites Dante to look in a certain direction by saying "drizza il nerbo del viso" (literally "direct the nerve of vision"), and the commentator recalls what has already been illustrated ("Come detto è poco avanti") about the nerve that moves from the brain and enables the visual process ("il vedere procede da uno nerbo che si muove dal cerabro"). Virgil's exhortation also prompts Benvenuto to provide some details about the optic nerve ("in oculo est quidam nervus [...], vocatur opticus"). The master from Imola reiterates that the "nerbo" means intellectual acumen ("robur intellectualis oculi" and "acumen intellectus, qui est oculus animae").[41] Furthermore, he points out that, according to natural scientists ("ut physici tradunt"), this nerve acts as a channel conveying images from the pupil to the brain ("transmittit species rerum visarum ad intellectum, et respondet pupillae"). It is wider than any other nerve ("grossior nervus quam sit in toto corpore"), as can be seen from the head of a kid or other animal ("sicut tu potes videre in capite unius capretti vel alterius animalis").[42] A slight different view of the "nerbo" is expressed by Guglielmo di Maramauro, who specifies that the optic nerve ("se chiama 'otiquo'") has the task of turning the eye in the direction of that which one wishes to look at ("el qual ne fa gitar l'ochio da ogni lato donde l'omo vol riguardare").[43]

Another passage that requires Dante's readers to consider the anatomy of the eye is the elaborate simile that occurs in *Par.* XXVI, 70–79:

> E come a lume acuto si disonna
> per lo spirto visivo che ricorre
> a lo splendor che va di gonna in gonna,
>
> e lo svegliato ciò che vede aborre,
> sì nescïa è la sùbita vigilia
> fin che la stimativa non soccorre;
>
> così de li occhi miei ogni quisquilia
> fugò Beatrice col raggio d'i suoi,
> che rifulgea da più di mille milia:
>
> onde mei che dinanzi vidi poi[44]
> (*Par.* XXVI, 70–79)
>
> [As sleep is broken by a piercing light / when the spirit of sight runs to meet the brightness / that passes through its filmy membranes, / and the awakened man recoils from what he sees, / his senses stunned in that abrupt awakening / until his judgment rushes to his aid / exactly thus did Beatrice drive away each mote / from my eyes with the radiance of her own, / which could be seen a thousand miles away, / so that I then saw better than I had before].

The poet compares his hazy vision, later healed by the rays coming from Beatrice, to the bewilderment of someone who is suddenly awakened by a bright light. In this parallel, Dante depicts the movement of the visual spirit as it approaches the splendour that penetrates through the various membranes of the eye. To comment on these verses, Iacomo della Lana dwells on the anatomy of the eye ("la notomia de i ochi"), explaining that the visual organ is composed of multiple sheaths ("l'ochio ha cumposto de pluxur tongeghe"), called "gonna" by the poet.[45] These membranes are one on top of each ("l'una sovra l'altra") and their centre is that point in the pupil where visual power properly lies ("nel centro hano solo quel punto in la pupilla dove è lo numero, c'ha la vertú visiva, sì termena").[46] After passing through the air and the various sheaths, starting from outermost one ("primo apar nella extrinseca, che termena cum l'aire"), the image ("spetia visibelle") reaches the centre of the eye ("fino al centro dell'ochio"). There it multiplies ("si multiplica"), as Aristotle says in the second book of *De anima* ("sí come apare in secundo *De anima* dal Philosopho"), and is then mirrored by being 'delivered' to the *sensus communis* ("s'aspieca nel senso comune").

The Anonimo Fiorentino repeats almost verbatim what he reads in Lana. Benvenuto merely refers to the teaching of experts in the field ("est notandum quod sicut tradunt physici oculus habet tunicas etc."), while the *Ottimo* offers a rather detailed description.[47] While noting that it would be long and unfruitful to cover the topic ("ed a tractare [...] sarebbe lungo ordino e con poco frutto"), he mentions the names of the seven sheaths ("sette tuniche" and "tunica viene a dire gonnella") of the eye, except for the *coniunctiva.* First comes the "retina", which "arises from veins and arteries and is woven like a web" ("nasce dalle vene e da l'arterie e in modo di rete è contessuta"). Then come the "secondina", the "sclirotica", and the "tunica aranea", which lies between the "cristallino" and the "albugineo umore" (i.e. aqueous humour); it is through this last sheath the visual spirit penetrates ("per la quale passa lo spirito della veduta"). Last come the "uvea" and the "cornea". In his conclusion, the *Ottimo* also makes his source explicit, namely Bartholomew the Englishman's encyclopedia: "In libro quinto *De proprietatibus rerum* habetur, capitolo iiii".

Galen's name ("ait sic Galienus") is instead recalled by Pietro Alighieri, who says that "our seeing is in the crystalline humour" ("nostrum videre est in humore cristallino"), which is separated from the cornea by another membrane ("inter quem et tunicam corneam interponitur unica tunica").[48] Behind this "tunica" (the uvea) there is the aqueous humour ("humor albugineus"), whereas between it and the "cristellinum" there lies the "tela aranea", which has a hole, the pupil, through which the visual spirit passes ("cuius foramen est pupilla").[49] Finally, Francesco da Buti more generically mentions natural philosophers ("diceno li Naturali") and compares the *tunicae* to various layers of leaves ("l'occhio è composto di più sode toniche come foglie"), in the middle of which ("in mezzo di quelle, sì come nel centro") is a liquid where the visual power lies ("è un umore in che sta la virtù visiva tra foglia e foglia").[50]

In sum, the anatomical descriptions examined here provide no original insights (nor should we expect them to). However, they may be of interest in understanding what sources Dante's commentators used and what notions they deemed fit to share with a wider readership. Moreover, these texts, particularly those in the vernacular, deserve attention from a lexical point of view, as they allow us to verify how certain technical terms and theories were circulated and translated in a context where Latin was not the prevailing language.[51]

5 Eye-Deceiving Conditions and Circumstances

As we have already seen, for example with regard to Farinata's answer, Dante and his commentators are well aware that the distance and the characteristics of the *medium* or "mezzo" – that is, the element that lies between the eye and the object and may be more or less terse and bright – significantly influence visual perception. As Andrea Lancia puts it, "the author observes that the sense of the eye is deceived by distance in space or darkness in the air".[52] Not unexpectedly, this remark is made with reference to *Inf.* XXXI, 19. Almost at the end of the first *cantica*, much of this canto builds on the difficulties due to the gloom and misunderstandings, which gradually melt away, giving way to fear, as the poet and Virgil approach the towers/giants.

Right from the start (vv. 10–11) Dante creates a twilight atmosphere, with little chance to catch a glimpse of what lies ahead:

> Quiv'era men che notte e men che giorno,
> sì che 'l viso m'andava innanzi poco;
> ma io senti' sonare un alto corno,
>
> tanto ch'avrebbe ogne tuon fatto fioco,
> che, contra sé la sua via seguitando,
> dirizzò li occhi miei tutti ad un loco.
> (*Inf.* XXXI, 10–15)

> [Here it was less than night and less than day – / I could not see too far ahead. / But I heard a horn-blast that would have made / the loudest thunderclap seem faint. / To find its source I turned my eyes / back to the place from which the din had come].

It is then the sound of a horn, more terrible than that of Roland's olifant, that directs the poet's gaze (v. 15) towards the supposed towers, and makes him ask Virgil for enlightenment as to their whereabouts. The guide replies:

> Ed elli a me: "Però che tu trascorri
> per le tenebre troppo da la lungi,
> avvien che poi nel maginare abborri.
>
> Tu vedrai ben, se tu là ti congiungi,
> quanto 'l senso s'inganna di lontano;
> però alquanto più te stesso pungi".
> (*Inf.* XXXI, 22–27)

> [And he to me: "Because you try to pierce / the darkness from too far away, / it follows that you err in your perception. / When you are

nearer, you will understand / how much your eyesight is deceived by distance. / Therefore, push yourself a little harder"].

Since he looks too far into the distance and almost in the absence of light, Dante is confused and cannot form a mental image that matches the truth. Getting closer he will instead be able to peer through the thick, dark blanket (v. 37: "forando l'aura grossa e scura") and realise that he is facing frightening giants. A beautiful comparison with fog, which, as it dissipates, allows what was previously concealed to emerge, completes the description of this gradual unveiling:

> Come quando la nebbia si dissipa,
> lo sguardo a poco a poco raffigura
> ciò che cela 'l vapor che l'aere stipa.
> (*Inf.* XXXI, 34–36)

> [As, when the mist is lifting, little by little we discern things hidden in the air made thick by fog].

Another contribution in this volume focuses on the optical illusion the poet sketches at the end of the canto (vv. 136–141), when he compares Antaeus, in the act of stooping, to the Garisenda tower in Bologna, which gives the impression of falling on the observer if the clouds move in the opposite direction to its inclination.[53] The optical phenomenon on which the simile is based is rather neglected by early commentators, who were more interested in providing details about the building, its history, Dante's stay in Bologna, etc. Only Iacomo says that the tower seems to be falling ("el par puro ch'ella caça") because of the visual ray that penetrates deep into the air ("sì se interna lo raggio visuale in l'aere").[54]

More attention is paid to the atmosphere in which the scene is set and to the problem of the distance between the eye and the thing to be seen. The *Ottimo* remarks that the gloomy air between Dante's pupil and the giants ("l'aere tenebroso, ch'era mezzo tra le pupilla de li occhi e l'obietto, ch'erano li giganti") does not make things look the way they are ("non lasciava fare vera aprensiva") and therefore induces misjudgment ("faceva falso iuditio").[55] Benvenuto also takes into account darkness and the position of the object in space ("propter distantiam medii et maxime in tenebra") and provides a second example of deception that frequently occurs, namely mistaking a man for a tree ("saepe capit hominem pro arbore").[56] It is, however, Francesco da Buti who offers a more comprehensive explanation.[57] The commentator makes it clear that Dante cannot recognise what is in front of him because he is in a dimly lit place and looking from too far away ("Perchè t'è troppo di lungi la vista, conviene correre più che non può nel luogo tenebroso"). Then he wonders what sight might perceive if the place were not dark but bright ("imperò che cosa potrebbe la vista nel luogo chiaro, che non può nel luogo tenebroso?"), and this question is followed by some specifications. Francesco points out that in the science of vision ("nella prospettiva"), as in other circumstances ("come nell'altre cose"), there must be a proportion ("si richeggono proporzioni") among the elements involved. Visual power ("virtù visiva") can be more or less effective depending on its capacity ("la sua potenzia"), distance ("la distanzia del luogo"), the object's size ("la quantità dell'obietto"), and brightness ("la chiarezza della luce"). Consequently, the same thing at a certain distance is seen differently by different eyes, depending on how powerful they are ("secondo che la virtù visiva è maggiore in uno che in un altro"); furthermore, the same eye sees different by day and by night ("altrimenti si comprende la cosa di dì, altrimenti di notte"), as well as from a distance and up close ("altrimenti da presso, altrimenti da lungi").

The Anonimo Fiorentino repeats roughly the same content, but includes a noteworthy detail about the density of the *medium*.[58] The human eye, he explains, may not see well and take one thing for another because the distance is either too small or too great ("secondo la distanzia o picciola o grande"), and because of the qualities of the "mezzo". After saying that one can see better from afar if the air is bright ("per l'aria chiara si vede più da lungi che per la oscura"), he dwells on an optical phenomenon that is not strictly related to Dante's verses and concerns the density of the "mezzo", which can be either thick or thin ("grosso o sottile"). Indeed, the Anonimo clarifies that a thing surrounded by air or another 'thin' substance ("per l'aria sottile et per la materia sottile") appears smaller in size and, conversely, appears larger if it is in something dense ("nella materia grossa appare maggiore"). As a consequence, since water is denser than air ("l'acqua è di più grossa materia che non è l'aere"), a coin in a glass of water looks bigger than outside it ("mettendo uno danaro nell'acqua, il danaro parrà maggiore che di fuori dell'acqua").[59]

The note about the quality of the *medium* seems to be unprecedented among commentators, while on several occasions, besides those already recalled, Dante's readers stress the fact that proportion plays an important role in visual perception (and, therefore, in scientific reasoning about it). Iacomo della Lana, for example, explicitly mentions optical science ("como apare in perspetiva") in relation to those verses in which Dante, almost at the top of Purgatory, admires the stars, which seem brighter and larger than usual, that is, on earth (*Purg.* XXVII, 89–90: "vedea io le stelle / di lor solere e più chiare e maggiori"). In this respect, Iacomo notes that "the closer man comes to a shiny, luminous body, the greater that body appears in size and luminous intensity".[60] Similarly, with regard to Dante's failure to recognise the candelabra in *Purg.* XXIX, the commentator reiterates – once again referring to the *scientia perspectiva* ("sí come chiaro appare nella prospetiva") – that distance alters perception and can mislead the sense of sight ("lo senso del viso s'inganna"), insofar as "the space should be proportionate to the visual power and to the thing to be seen".[61] Iacomo thus stresses proportions, while Francesco da Buti, as seen above, makes a more complex argument and compares two different theories of vision. When commenting on this passage, the *Ottimo* and Benvenuto add nothing relevant from the point of view of optics, but the latter inserts a clarification about the phrasing and the grammatical subject ("Et vide quod in constructu debes incipere a longo tractu *etc.*").[62]

Also with regard to the poet's gaze in *Par*. XXVII, which follows the luminous spirits upward as long as it can during the extraordinary snowfall, Iacomo emphasises the need for proportion between the size of the object and the space separating it from the eye ("lo meço convene esser proportionado tra la cosa che se vede e 'l vixo").[63] He begins there by quoting Aristotle's *De anima* ("sí cum' dise 'l Philosopho nel secundo *De anima*") and notes that a small thing slips out of sight ("se perde ad ochio") more easily than a large one, particularly if the distance between it and the observer is beyond proportion ("quando lo meço è sí grande che ceda la proportione della cosa"). This, he concludes, is the reason why the saints, having strayed beyond measure ("cedea lo meço a la proportione") into the bright air ("l'aire luminoso"), can no longer be followed by the poet's eye ("lli perdé ello ad ochio").

Although it is not possible to develop the discussion any further here, to the excerpts examined we could add others in which Dante's commentators underline that distance also changes the perception of the shapes of objects. Concerning the river of light that becomes a lake, and then the white rose at the end of Paradise, Iacomo della Lana proposes an accurate geometric demonstration ("Sia linea AB lo extremo della riva de' fiuri da l'una delle parti de fora, etc.").[64] Francesco da Buti in turn explains that one realises the actual shape of a round thing of ample size ("una cosa tonda [...] molto ampia") only when approaching it ("come l'uomo s'appressa ad essa") or looking at it from above ("o se l'omo sallie in alto"), whereas before that it appears to be a kind of line ("parrà lunga").[65]

6 Conclusions: First Results and Next Steps

The review of passages presented here clearly shows that early commentators grasp the centrality of light and vision in the *Commedia*. At the same time, these readers of Dante prove that they know the basics of optics and, albeit to varying degrees, are aware of contemporary debates about the functioning of sight, light propagation, etc. A few developments in this research are on the horizon. First, there are other verses from the three *cantica* and excerpts from the commentaries to consider, particularly those dealing with the movement of light rays and mirrors – a couple of examples will be given shortly. In addition, it would be worthwhile to investigate the sources explicitly cited or employed without naming names, and to draw a comparison with what Dante says about optical matters in the *Convivio*. Finally, the vernacular commentaries might be a valuable field of inquiry to gain insights into the lexicon, by studying how the technical terms of a discipline that originated in Latin are rendered in the Italian of the time. Advancement along this path will, on the one hand, enable an increasingly accurate contextualisation of Dante's work with reference to one of its fundamental themes. On the other hand, a substantial pool of texts – hitherto almost unexplored from this perspective – is being brought to the attention of scholars of the history of optics, and they may prove valuable resources for the study of the popularisation of this discipline outside the proper scientific context. Indeed, Dante's commentators surely drew on those "popular conduits" through which the "visual paradigm filtered down from scholastic circles to a broader, less formally educated community". At the same time, they contributed to spreading "optical literacy", that is, to the dissemination (and vernacularisation) of the science of light and vision.[66]

The discussion developed so far can, therefore, be brought to a close with a final example that can also be regarded as an opening to future 'explorations'. In *Inf.* XXIII, 25–27 Virgil invokes a comparison with mirrors to say that Dante's thoughts and feelings are immediately reflected in his mind:

> E quei: "S'i' fossi di piombato vetro,
> l'imagine di fuor tua non trarrei
> più tosto a me, che quella dentro 'mpetro".
> (*Inf.* XXIII, 25–27)
>
> [And he: "If I were made of leaded glass / I could not reflect your outward likeness / in less time than I grasp the one inside you"].

The image used by the poet prompts Andrea Lancia to provide some clarifications on mirrors and their properties ("Speculum et eius proprietas").[67] He explains that a mirror is a leaded glass ("vetro impiombato") that, in the poet's words, draws the image to itself, as a magnet does with iron ("traela a sé, come calamita ferro"). Thus, the commentator points out, Dante settles here a specific issue ("qui solve una questione") regarding the way in which a mirror works, namely whether it is the image that reaches the reflecting surface ("se l'ymagine dello specchiante va allo specchio") or the surface that draws the image towards itself ("o lo specchio trae a sé l'ymagine dello specchiante"), arguing in favour of the latter explanation ("e dice che 'l trae, non va a llui").

The Anonimo Fiorentino also dwells on this matter at length.[68] First he emphasises that the technical knowledge involved, that is, optical science ("quest'arte prospettiva"), is uncommon and complex ("è poco in uso, et forte è malagevole"). As for the specific question of how images reach mirrors, there would not appear to be any solution at hand, since no philosopher makes a clear statement on this ("chiaramente per veruno filosofo si disfinisce"). At least it can be taken for granted ("se non che dicono etc.") that the image of each thing arrives in a straight line at the mirror placed in front of the object ("per retta linea ferisce all'opposito suo nello specchio").

Therefore, images cross thin and transparent surfaces – and they could do the same with a clear liquid ("et in qualunche altra materia che sia rara et lucida, ovvero liquida") – and stop where they find something thick and dark ("passa infino alla cosa densa et oscura"), for example the leaded part of a mirror; they are then reflected backward and return to the eye ("ritorna addirietro, et manifestasi agli occhi nostri").

The commentator also wants to show what can be achieved through the skilful use of mirrors. He maintains that, by making use of a few reflective surfaces, one can instantaneously see what is happening in a place many miles away ("di specchio in specchio in uno stante si possono vedere le cose che universalmente si fanno di lungi molte miglia"). He also seems to report an experiment that was actually carried out ("dicesi che fare si poterebbe, e fatto è stato"). He explains that through a system of three mirrors correctly arranged ("per quello modo che 'l maestro saprebbe acconciare") on top of as many mountains, from Monte Morello, the highest peak in the Florentine plain, one could see in real time what is happening in Bologna ("si vederebbe dallo specchio di Monte Morello ciò che fatto fosse in Bologna in uno stanti"). Finally, the Anonimo reiterates that those who have experienced this know how such a result can be achieved ("come questo possa essere chi ha fatta la sperienza il sa").[69] Once again, then, Dante's ingenious similes – built on optical phenomena frequently encountered in everyday life, but whose explanation was unknown to most people – led contemporary interpreters to investigate certain matters, to gather and convey information, and to digress about topics close to – and sometimes distant from – the starting idea. And along with them we also discover something new about the poet's world and words, and about the notions and uses of optical science and tools in 13th- and 14th-century Italy.

Notes

1 This contribution is part of a research project that is funded by the Swiss National Science Foundation and is currently being conducted at the University of Zurich (Ambizione grant, PZ00P1_201935).
2 See Patrick Boyde's preface to Gilson, *Medieval Optics*, XIII. See also Piccolino, "The Visions of Dante", in this volume.
3 The bibliography on this subject is huge. See Smith, *From Sight to Light*, where many other references (including the pioneering studies by David C. Lindberg and Graziella Federici Vescovini) can be found.
4 Much has been published on the topic. Among other publications (and in addition to Gilson's monograph), see Akbari, *Seeing through the Veil*; Ariani, *Lux inaccessibilis*; Panti, "Il visibile 'per retta linea'".
5 See Gilson, "Optics in and through Dante"; the author briefly deals with Trecento commentators at 505.
6 Needless to say, the commentators' level of education is higher than that of the general public. Albeit in relation to a later period, notable general considerations on the study of Dante's reception can be found in Gilson, *Leggere Dante*, 19–44.
7 See Galli, "e però puote anche la stella parere turbata". See also Ledda, "Filosofia e ottica".

8 For a general introduction (date, authorship, etc.) to commentaries transmitted in manuscripts and dated to before 1480, see Malato and Mazzucchi (eds.), *Censimento dei commenti*.

9 Initial surveys were conducted through the DDP: "Dartmouth Dante Project", https://dante.dartmouth.edu/ (accessed on 3 May 2023). Whenever feasible, commentaries have been read and quoted on the basis of their respective critical editions. When no critical editions are available, the texts have been quoted from 19th-century editions, even though these publications are not always reliable (especially that of the Anonimo Fiorentino's commentary). As for Benvenuto da Imola's commentary, comparisons have been made with the editions of his *Lectura Bononiensis* (1375) and the *Lectura Ferrariensis* (1375–1376). I would like to thank Luca Azzetta for his suggestions on the editions of early commentaries and the valuable materials he shared with me. Of course, any errors or inaccuracies are my responsibility.

10 Iacomo's commentary has been published by Volpi as a synoptic edition of two different versions, in the Bolognese vernacular and in the Tuscan vernacular respectively. In the table I have quoted both, while in what follows I will refer to the Bolognese version of Iacomo's text, and I thank the editor for his suggestions in this regard.

11 See Dante, *Convivio*, 234. Benvenuto da Imola, *Comentum*, vol. 3, 406: "prospectiva, quae partim spectat ad philosophiam, partim ad geometriam". The commentator explains the term "arte" in *Purg*. XV, 21.

12 Iacomo della Lana, *Commento*, vol. 2, 1652 (*Purg*. XXXIII, 103): "sí cum' chiaro apare nella *Perspetiva* in lo capitulo della diversità de l'aspetto in li canoni sovra le Tavole Tolletane".

13 Iacomo della Lana, *Commento*, vol. 1, 852 (*Inf*. XXXI, 10).

14 The *Commedia* is always quoted from the edition by Chiavacci Leonardi. The English translation is the one by Hollander, available on the "Princeton Dante Project" website, https://dante.princeton.edu/pdp/ (accessed on 28 April 2023).

15 Dante Alighieri, *La Divina Commedia*, vol. 1, 324. Moreover, Nicola Fosca wonders whether Farinata's "noi" refers to all the damned or just to the Epicureans: see Fosca's commentary (2003–2015) in the DDP (*Inf*. X, 100–105).

16 Singleton, "Commentary", quoted from the DDP (*Inf*. X, 100).

17 Iacomo della Lana, *Commento*, vol. 1, 344 (*Inf*. X, 100).

18 *Ottimo commento*, vol. 1, 252 (*Inf*. X, 100–108); Amico dell'Ottimo, *Chiose*, 102 (*Inf*. X, 100); see also 754.

19 The quotation is from the second redaction of Pietro's commentary (1342–1355): Pietro Alighieri, *Comentum*, 297 (*Inf*. X, 94–114). The 'thickening' of the *virtus visiva*, which makes one unable to see farther than a span, is also mentioned by Iacomo della Lana and the *Ottimo* when commenting on *Par*. XIX, 79. This canto is replete with references to vision, but there is not much optical material in the commentaries.

20 Comment on *Inf*. X, 100: "coloro c'ànno la mala luce sono naturalmente i vecchi che vegono le cose da lungi meglio che da presso." The commentary, known as *Chiose cagliaritane* and dated to the late 14th century, is quoted from the DDP (ed. by Carrara).

21 Boccaccio's commentary is always quoted from the DDP (ed. by Padoan).

22 Part of the quoted passage is transmitted by manuscripts E (Estense) and S (Strozziani) and is only included by the 19th-century editor in the critical apparatus. The integration in E and S (or a similar one) seems essential, however, to complete the meaning of the sentence: "nam quidam habent visum multum sed non clarum, et tales vident melius a longe <quia radii visuales egrediendo serenantur: alii autem habent visum clarum, sed non multum>, et tales non vident nisi a propinquo. Modo autor hic loquitur de illis qui vident rem a longe, sed non de

prope". See Benvenuto da Imola, *Comentum*, vol. 1, 352. What we read in Benvenuto's two *Lecturae* is not helpful in this regard.

23 Here the author pays much attention to the theological question of the foreknowledge of the damned and reports various opinions. According to some, the inhabitants of hell do not see anything ("illi qui habent malam lucem non vident aliqua perfecte") unless there is an external intervention ("nisi secundum quod ostenditur sibi ab aliis qui ostendunt sibi rem"); therefore, they know neither the present nor the future, but can only be aware of something if they receive news from others (e.g. a demon or other spirit who has entered the infernal realm). This explanation is orthodox ("pulcra et catholica") but does not respect the text ("litera tamen non patitur eam"). Even less acceptable ("ista expositio est quaedam fuga ignorantiae") is the hypothesis that the damned can only foresee death, since it contradicts the preceding verses.

24 Amico dell'Ottimo, *Chiose*, 146 (*Inf.* XV, 13–16).

25 Before referring to the (pseudo)Aristotelian *Problemata*, Benvenuto describes how the elderly tailor narrows his eyes when threading a needle ("contrahit cilia, ut virtus visiva fiat fortior quando vult ponere filum in parvum foramen acus"). This act concentrates visual power into a smaller space, enhancing sharpness and effectiveness. ("quia virtus tunc magis unita exit per minorem viam"). The same source is mentioned by the commentator from Imola in relation to another gesture related to sight. In *Purg.* XV, 13–15, Dante shields his eyes with his hand in order to protect them from excessive light. Benvenuto again points out that the aforementioned Aristotelian work accounts for the phenomenon: "Unde Aristoteles libro Problematum circa principium quaerit propter quid versus solem opponentes manum ante lumen magis videmus. Et dicit quia superabundantia prohibita a manu visum non laedit." Francesco da Buti also dwells on these verses, provides the etymology of "solecchio" (it is a diminutive meaning "small sun": "è a dire picculo Sole"), and emphasises that the thing to be seen must be commensurate with one's visual capabilities ("contemperata a la vista"). Conversely, if it exceeds them ("avansa la potenzia visuale"), it must be scaled down ("per parificamento e reduzione del superfluo a parità et equalità de la porta visiva") by closing the eye or shielding it with the hand. See *Commento di Francesco da Buti*, vol. 2, 349 (*Purg.* XV, 1–15). Similar to Benvenuto, Giovan Battista Gelli, in his commentary (1559), makes reference to the *Problemata* (albeit addressing different questions from those recalled earlier) while discussing optical topics in relation to *Inf.* X, 100–108.This passage is cited by Gilson: Gilson, "Optics in and through Dante", 515–516.

26 *Commento alla Divina Commedia d'Anonimo Fiorentino*, vol. 1, 241–242 and 256–257.

27 *Commento di Francesco da Buti*, vol. 1, 285 (*Inf.* X, 61–72). Concerning Farinata's words, Francesco merely observes: "Et a questo risponde, che li passati veggono come colui ch'à mala luce, che vede le cose da lungi; ma non da presso: e così ellino veggono il futuro; ma non il presente"; 286 (*Inf.* X, 73–84).

28 The textual variant is reported in the apparatus and the name of Francesco da Buti is mentioned in the footnotes in Petrocchi's edition: *La Commedia secondo l'antica vulgata*, vol. 2, 166 (*Inf.* X, 69).

29 Francesco's mention of "prospettiva" would actually be a reference to a specific treatise on optics, if it is correct to capitalise the P as in the 19th-century edition; if not, it must be a more generic reference to the science of vision.

30 *Commento di Francesco da Buti*, vol. 2, 702–703 (*Purg.* XXIX, 43–57).

31 The illustration is given here as in the 19th-century edition (at 703), since I have not yet examined the manuscripts.

32 In other passages some commentators, following Aristotle, define colour as the proper object of the sense of sight. See, by way of example, Iacomo della Lana on

Purg. XVI: "lo color ch'è sempre oietto della vertù visiva"; Iacomo della Lana, *Commento*, vol. 2, 1250.

33 These theories conflicted less sharply than we tend to believe today; see the recent Lička, "The Visual Process".

34 *Ottimo commento*, vol. 3, 1364 (*Par.* III, 13–15).

35 *Ottimo commento*, vol. 3, 1906 (*Par.* XXXIII). On the multiplication of visual *species* in the pupil, see also 1695–1696 (*Par.* XXI, 16–18), where the commentator refers to Aristotle's *De anima* and *De sensu et sensatu*, and 1773 (*Par.* XXVI, 103–114).

36 *Commento alla Divina Commedia d'Anonimo Fiorentino*, vol. 1, 204. See also Dante's *Convivio*, III, ix, 6–16.

37 *Commento alla Divina Commedia d'Anonimo Fiorentino*, vol. 1, 204–206. The description by the Anonimo Fiorentino bears some similarities to Bacon's account of the anatomy of the eye in his *Perspectiva*: Lindberg, *Roger Bacon and the Origins*, 26–33 (p. I, dist. 2, ch. 2–3). Concerning the anatomy of the eye, its representations, and the specific terms used to describe it from antiquity to the late Middle Ages, see Raynaud, *Eye Representation*, 3–70, 477–486. See also Smith, *From Sight to Light*, 36–43, 184–188.

38 To be more precise, the cornea is the membrane, while the 'white' is the aqueous humour or *humor albugineus*. The Anonimo also adds something about the uses of horn in writing practices: "et è detta cornea da quello corno, cioè da quello osso che si pone alcuna volta inanzi a certe lettere di tavola o d'altro, acciò che le lettere si conservino molto et possansi leggere". As the editor points out, the commentator is probably thinking of 'horn windows', of thin, transparent horn plates applied to the covers of ancient codices to protect the title of the work. I once again wish to thank Marco Cursi for his valuable help with palaeographic and codicological issues.

39 As Smith explains, when commenting on Alhacen's description of the eye, "the uveal tunic continues beyond the circle created by the intersection of the sclera and the cornea, but it falls short of completion by the amount occupied by the circular opening of the pupil": Smith, *Alhacen's Theory of Visual Perception*, vol. 1, LVII–LVIII.

40 *Commento alla Divina Commedia d'Anonimo Fiorentino*, vol. 1, 230.

41 Benvenuto already includes some information about the optical nerve in the *Lectura Ferrarensis*: "*nerbo*, idest acumen: idest, respondet nervus qui ministrat visum. Iste est groscior nervus corpore, qui ministrat virtutem visivam, et respondet <in> medio occulo, in pupilla; atrahit res vivas exteriores et mittit in mentem; vocatur 'nervus oticus' penes phisicos." (and note the use of the verb 'to attract', "atrahit", which could imply an active role). See Benvenuto da Imola, *Lectura Dantis Ferrarensis*, 216 (*Inf.* IX, 73–75).

42 Benvenuto da Imola, *Comentum*, vol. 1, 319.

43 Guglielmo Maramauro's commentary is quoted from the DDP (ed. by Pisoni and Bellomo).

44 Concerning the meaning of "aborre" (v. 73) and two alternative interpretations of it, see Dante Alighieri, *La Divina Commedia*, vol. 3, 724.

45 Iacomo della Lana, *Commento*, vol. 4, 2444 (*Par.* XXVI, 72).

46 The meaning of the last part of the sentence quoted ("dov'è [...] termena") is unclear. The 19th-century edition, which can be consulted in the DDP, reads: "più toniche [...]. le quali nel centro hanno solo quello punto in la pupilla dove lo numero, che ha la virtù visiva, si termina," which could be taken to mean "where the number (of tunics) that has visual power ends". I wish to thank Mirko Volpi for the time he has spent discussing this passage with me, providing valuable suggestions.

47 *Commento alla Divina Commedia d'Anonimo Fiorentino*, vol. 3, 470; Benvenuto da Imola, *Comentum*, vol. 5, 378–379; *Ottimo commento*, vol. 3, 1772 (*Par.* XXVI, 70–78).

48 Pietro Alighieri, *Comentum*, vol. 2, 1156.
49 Usually, the pupil is considered to be a hole in the uvea. In the *Chiose Cassinesi* (ca. 1365) we find an identical explanation, while the one in the *Chiose Ambrosiane* (1355) is more detailed.
50 *Commento di Francesco da Buti*, vol. 3, 696 (Par. XXVI, 67–78).
51 Although Margherita Quaglino has mostly dealt with later authors (including Leonardo), I will here refer to her studies, including Quaglino, "Leonardo".
52 Andrea Lancia, *Chiose*, vol. 1, 453 (*Inf.* XXXI, 19): "nota l'autore come il senso de l'occhio per distantia di luogo o per tenebrositade d'aere s'inganna".
53 See Piccolino, "Moving Clouds", in this volume.
54 Iacomo della Lana, *Commento*, vol. 1, 860 (*Inf.* XXXI, 136).
55 *Ottimo commento*, vol. 1, 642 (*Inf.* XXXI, 22–27).
56 Benvenuto da Imola, *Comentum*, vol. 2, 458.
57 *Commento di Francesco da Buti*, vol. 1, 787 (Inf. XXXI, 19–27).
58 *Commento alla Divina Commedia d'Anonimo Fiorentino*, vol. 1, 653–654.
59 A number of optical illusions and phenomena, including the appearance of a coin in a glass of water, will be revisited a century later in Varchi's exposition on Dante's poem. See Gilson, "Optics in and through Dante", 514.
60 Iacomo della Lana, *Commento*, vol. 2, 1510 (*Purg.* XXVII, 89): "quanto l'omo s'avisina più al corpo lucido e luminoso, tanto apare ello maore sí in quantitade come in qualitade".
61 Iacomo della Lana, *Commento*, vol. 2, 1560 (*Purg.* XXIX, 46): "'l spatio convene essere proportionado secundo la vertú visiva de quel che vede e segondo la cosa ch'è veçuda."
62 Benvenuto da Imola, *Comentum*, vol. 4, 458.
63 Iacomo della Lana, *Commento*, vol. 4, 2478 (*Par.* XXVII, 74).
64 Iacomo della Lana, *Commento*, vol. 4, 2598 (*Par.* XXXI).
65 *Commento di Francesco da Buti*, vol. 3, 795 (Par. XXXI, 82–96).
66 Smith, *From Sight to Light*, 280 and 287.
67 Andrea Lancia, *Chiose*, vol. 1, 363–364 (*Inf.* XXIII, 25).
68 *Commento alla Divina Commedia d'Anonimo Fiorentino*, vol. 1, 492–493.
69 Further questions and observations about reflecting surfaces can be found in the comments on *Purg.* XXV, 25–27, where Virgil gives the example of movements being reflected in a mirror with the aim of explaining the relationship between souls and aerial bodies to Dante. See, for example, the *Ottimo*: *Ottimo commento*, vol. 2, 1203 (*Purg.* XXV, 25–27).

Bibliography

Primary Sources

Amico dell'Ottimo. *Chiose sopra la 'Comedia'*. Edited by Ciro Perna. Rome: Salerno, 2018.

Benvenuto da Imola. *Comentum super Dantis Aldigherij Comoediam*. Edited by Giacomo Filippo Lacaita, 5 vols. Florence: Barbèra, 1887.

Benvenuto da Imola. *Lectura Dantis Ferrarensis*. Edited by Carlo Paolazzi, Paolo Pasquino, and Fabio Sartorio. Ravenna: Longo, 2021.

Commento alla Divina Commedia d'Anonimo Fiorentino del secolo XIV. Edited by Pietro Fanfani, 3 vols. Bologna: Gaetano Romagnoli, 1866–1874.

Dante Alighieri. *La Commedia secondo l'antica vulgata*. Edited by Giorgio Petrocchi, 4 vols. Florence: Le lettere, 1966–1967.

Dante Alighieri. *The Divine Comedy*. Translated, with a commentary, by Charles S. Singleton. Princeton, NJ: Princeton University Press, 1970–1975.

Dante Alighieri. *La Divina Commedia*. Edited by Anna Maria Chiavacci Leonardi, 3 vols. Milan: Mondadori, 2005.

Dante Alighieri. *Convivio*. Edited by Gianfranco Fioravanti and Claudio Giunta. Milan: Mondadori, 2019.

Francesco da Buti. *Commento di Francesco da Buti sopra La Divina Commedia di Dante Allighieri*. Edited by Crescentino Giannini, 3 vols. Pisa: Fratelli Nistri, 1858–1862.

Giovan Battista Gelli. *Commento edito e inedito sopra la Divina Commedia*, a cura di Carlo Negroni. Florence: Bocca, 1887.

Giovanni Boccaccio. *Esposizioni sopra la Comedia di Dante*. Edited by Giorgio Padoan. Milan: Mondadori, 1965.

Guglielmo Maramauro. *Expositione sopra l' "Inferno" di Dante Alighieri*. Edited by Giacomo Pisoni and Saverio Bellomo. Padua: Antenore, 1998.

Iacomo della Lana. *Commento alla 'Commedia'*. Edited by Mirko Volpi and Arianna Terzi, 4 vols. Rome: Salerno, 2009.

Le Chiose Cagliaritane. Edited by Enrico Carrara. Città di Castello: Lapi, 1902.

Ottimo commento alla 'Commedia'. Edited by Giovanni Battista Boccardo (*Inferno*), Massimiliano Corrado *(Purgatorio)*, and Vittorio Celotto (*Paradiso*), 3 vols. Rome: Salerno, 2018.

Pietro Alighieri. *Comentum. Redazione ashburnhamiano-barberiniana*. Edited by Giuseppe Alvino, 2 vols. Rome: Salerno, 2021.

Secondary Sources

Akbari, Suzanne Conklin. *Seeing through the Veil. Optical Theory and Medieval Allegory*. Toronto, Buffalo and London: University of Toronto Press, 2004.

Ariani, Marco. *Lux inaccessibilis*. Rome: Aracne, 2010.

Galli, Francesca. "'e però puote anche la stella parere turbata' (*Conv.* III, IX 14). La percezione visiva fra teoria, vissuto e rappresentazione". In *L'esperienza di Dante*, edited by Giuseppe Ledda, 13–35. Ravenna: Centro dantesco dei frati minori conventuali, 2023.

Gilson, Simon A. *Medieval Optics and Theories of Light in the Works of Dante*. Lewiston, Queenston and Lampeter: The Edwin Mellen Press, 2000.

Gilson, Simon A. *Leggere Dante a Firenze*. Edited by Anna Pegoretti, translated by Giulia Gaimari. Rome: Carocci, 2019.

Gilson, Simon A. "Optics in and through Dante: Notes on Visual and Optical Theory in the Lezioni on Dante by Cosimo Bartoli and Benedetto Varchi". *Letteratura italiana antica* 20 (2019): 501–516.

Ledda, Giuseppe. "Filosofia e ottica nella predicazione medievale". In *Letteratura in forma di sermone*, edited by Ginetta Auzzas, Giovanni Baffetti, and Carlo Delcorno, 53–78. Florence: Leo S. Olschki, 2003.

Lička, Lukáš. "The Visual Process: Immediate or Successive? Approaches to the Extramission Postulate in 13th Century Theories of Vision". In *Medieval Perceptual Puzzles: Theories of Sense Perception in the 13th and 14th Centuries*, edited by Elena Băltuță, 73–110. Boston, MA and Leiden: Brill, 2020.

Lindberg, David C. *Roger Bacon and the Origins of Perspectiva in the Middle Ages*. Oxford: Clarendon Press, 1996.

Malato, Enrico and Mazzucchi, Andrea. Eds. *Censimento dei commenti danteschi. 1. I commenti di tradizione manoscritta (fino al 1480)*, 2 vols. Rome: Salerno, 2011.

Panti, Cecilia. "Il visibile 'per retta linea': Dante, la perspettiva e una proposta di emendamento in *Convivio* III IX 6". In *Dante e il mondo, tra realtà e poesia, tra storia e letteratura. Atti del LVIII Convegno storico internazionale. Todi, 10-12 ottobre 2021*, 143–180. Spoleto: Centro Italiano di Studi sull'Alto Medioevo, 2022.

Quaglino, Margherita. "Leonardo 'trattatore della luce.' Prime osservazioni sul lessico dell'ottica nei codici di Francia". *Studi di lessicografia italiana* 30 (2013): 93–132.

Raynaud, Dominique. *Eye Representation and Ocular Terminology from Antiquity to Helmholtz*. Amsterdam: Wayenborgh, 2020.

Singleton, Charles S. "Commentary" to Dante Alighieri. *The Divine Comedy*. Translated with a commentary by Charles S. Singleton. Princeton, NJ: Princeton University Press, 1970–1975.

Smith, A. Mark. *Alhacen's Theory of Visual Perception: A Critical Edition, with English Translation and Commentary, of the First Three Books of Alhacen's De aspectibus, the Medieval Latin Version of Ibn al-Haytham's Kitab al-Manazir*, 2 vols. Philadelphia, PA: American Philosophical Society, 2001.

Smith, A. Mark. *From Sight to Light. The Passage from Ancient to Modern Optics*. Chicago, IL and London: University of Chicago Press, 2015.

Acknowledgement

We acknowledge support for the Open Access publication of Chapter 7 has been funded by the Swiss National Science Foundation.

Schweizerischer Nationalfonds

8 A Medical Commentary on the *signa amoris* in the *Vita Nuova*

Francesco Brigo[1]

1 Introduction

Any neophyte reader approaching Dante's *Vita Nuova* is struck by the accuracy with which the poet recounts the physical experience of his first encounter with the beloved woman. It is probably the first time in Italian vernacular literature that a poet describes at length and in such detail the signs of love (*signa amoris*), with ample reference to contemporary medical theories. However, the interrelation between medicine and love literature that appears so distinctive in the *Vita Nuova* is far from being something unique and is instead emblematic of the circular knowledge system, rich in mutual influences between fields that today would be considered irreconcilable and totally distinct, so characteristic of the Middle Ages.[2]

The study of the almost osmotic relationship between medicine and love literature has given rise to a branch of textual criticism that has led to relevant contributions.[3] Among other achievements, this new field of research has revealed how certain metaphors and literary expressions should not be considered simply as rhetorical devices or topoi but should be traced back to a specific medical lexicon. All this has led to a renewed interpretative freshness, shedding new light on literary texts that seemed to have exhausted their hermeneutic potential.

The main methodological problem of this line of study is to establish with sufficient accuracy the direct knowledge of medical texts by poets and literary authors in the Middle Ages. A problem that results in the difficulty of ascertaining possession, circulation, and access to manuscripts. A problem that is even greater with Dante, since the manner in which he acquired his medical knowledge is still surrounded by more questions than answers. Beyond his membership in the Guild of Physicians and Apothecaries (in 1295) and the historic depiction of Dante dressed in a red *lucco*, the robe with the colour traditionally associated with the medical profession, there is little evidence of direct medical knowledge on his part. However, even if he never followed a systematic course of study in medicine, he was certainly well-versed in the subject, as evidenced by the numerous references to contemporary medical ideas scattered throughout his works.[4] In addition to specialised treatises,

DOI: 10.4324/9781003412601-9

in the Middle Ages, medical theories could also be transmitted through non-medical writings, such as encyclopaedias, scholastic texts, and philosophical and theological works.

This chapter tries to deal with the love described by Dante in the *Vita Nuova* as a medical commentary of his time – conducted *scientifico modo et veridico, tracto ex preceptis scientie naturalis* – would have done.[5] I focused mainly on the *signa amoris* that can affect the body (*signa ex parte corporis*) and the soul (*signa ex parte animae*), and their underlying mechanisms, which were discussed according to the medical theories of Dante's time.[6] The lack of sufficient detail prevented me from discussing love symptoms in anatomical order (*a capite ad calcem*) as a medieval treatise would have done.

Aware of the difficulties and uncertainties related to direct knowledge of medical texts by Dante, I tried to rely on the principle of plausibility by referring to contemporary writings that had relatively wide circulation and/or accessibility.

2 Love: A Matter of Spirits

Dante recounts his first encounter with Beatrice in a rather unusual way, describing how this marvellous apparition affected his body and soul:

> At that very moment, and I speak the truth, the vital spirit, the one that dwells in the most secret chamber of the heart, began to tremble so violently that even the most minute veins of my body were strangely affected; and trembling, it spoke these words: *Ecce deus fortior me, qui veniens dominabitur michi* [Here is a god stronger than I who comes to rule over me]. At that point the animal spirit, the one abiding in the high chamber to which all the senses bring their perceptions, was stricken with amazement and, speaking directly to the spirits of sight, said these words: *Apparuit iam beatitudo vestra* [Now your bliss has appeared]. At that point the natural spirit, the one dwelling in that part where our food is digested, began to weep, and weeping said these words: *Heu miser, quia frequenter impeditus ero deinceps!* [Oh, wretched me! for I shall be disturbed from now on].[7]

The poet stages a kind of theatrical performance enacted by various spirits that speak Latin, the language used in contemporary medical treatises, but adopting a quasi-biblical tone. Concealed beneath the rhetorical device of personification, the description is very precise and relies heavily on contemporary medical theories dating back to Claudius Galen.[8] According to the ancient Roman physician, three types of spirits governed the body, and each had a seat in a specific organ.[9] Digested food was turned into blood by the liver and then carried through the body in veins that contained the natural spirit (*spiritus naturalis*), responsible for nourishing the body's tissues. After reaching the heart, the natural spirit was refined by air from the lungs and became a vital spirit (*spiritus vitalis*), which was distributed to the body through the arteries to sustain and maintain life (*vita*).[10] It was also responsible for the emotions. See, for example:

> Allor fu la paura un poco queta,
> che nel lago del cor m'era durata
> la notte ch'i' passai con tanta pieta
> (*Inf.* I, 19–21)

> [Then the fear was somewhat quieted / that had continued in the lake of my heart / through the night I had passed so piteously].[11]

In a network of blood vessels at the base of the brain (the so-called *rete mirabile*, "wondrous net"), the vital spirit was transformed into the animal spirit (*spiritus animalis*, from *anima*, soul), responsible for sensation, movement, and higher functions (in humans). It was stored in the ventricles, a series of cavities in the brain that were thought to be hollow.[12] From there, the animal spirit could enter the nerves, conceived as a series of hollow channels, to subserve sensation and movement. Over the course of history, each ventricle ended up being associated with a specific function or power (*virtus*): initial processing of sensory and visual information in the anterior cavity, reasoning in the middle one, and memory in the posterior ventricle. This theory gained great popularity, partly due to the endorsement given by Saint Augustine, and became known as the cell doctrine (from *cella*, chamber) or theory of the inner senses.[13] It was refined by Avicenna, and adopted by Thomas Aquinas, Albertus Magnus, Roger Bacon, and Peter of Spain, among others.

From a literary perspective, the description of the physical effects of love mediated by personified spirits has its most immediate ascendancy in the works of Guido Cavalcanti. The rhetorical device of personification (*prosopopoeia*) is frequently used by this poet to refer to the spirits impaired by the sight of the beloved woman. Chapter XXV of the *Vita Nuova* deals with the appropriateness of using this expedient in poetry. Here, Dante refers to Guido as *mio primo amico* (my first friend) and recalls how they were well aware of the inappropriate use of prosopopoeia. As the following text shows, the description of the spirits addressing each other in *Vita Nuova* II is clearly recognised by Dante as belonging to this rhetorical device:

> So, if we find that the Latin poets addressed inanimate objects in their writings, as if these objects had sense and reason, or made them address each other, and that they did this not only with real things but also with unreal things (that is: they have said, concerning things that do not exist, that they speak, and they have said that many an accident in substance speaks as if it were a substance and human), then it is fitting that the vernacular poet do the same-not, of course, without some reason, but with a motive that later can be explained in prose.[14]

Thus, in the *dramatis personae* evoked by Dante in recounting Beatrice's first appearance, spirits seem to be endowed with prescience. In particular, the vital spirit acts as a forerunner of the god Love, announcing its arrival. Clearly, it is modelled on the biblical figure of John the Baptist, the forerunner of Christ, even in the words it speaks.[15] Notably, in the *Vita Nuova* (chapter XXIV), Giovanna – the woman loved by Guido Cavalcanti – is modelled on John the Baptist, as she prefigures and gives way to Beatrice. Taken together, these passages establish the following series of antinomies: Giovanna/Beatrice; John the Baptist/Christ; Guido Cavalcanti/Dante. In addition to the literal meaning, the reference to spirits in the first meeting with Beatrice appears functional to this underlying scheme.

3 Vital Spirit

In describing the bodily effects exerted by Beatrice's appearance, Dante begins with the vital spirit. Even from the succinct and simplified summary given above, it is evident that in his description, Dante does not follow the *ordo naturalis*. The rhetorical device of anaphora (*in quello punto*) is employed to emphasise the simultaneity of the impairment of spirits after the first encounter with Beatrice. Although this synchronous impairment would have been considered implausible by medieval medical theories, it reveals how this event profoundly affected the wholeness of Dante's body and soul. The *ordo artificialis* chosen by the poet to describe the effects on his spirits follows a precise hierarchy, with the greatest emphasis on the vital spirit and its seat, the heart, where Love will also reside ("Amore e 'l cor gentil sono una cosa; Amor per sire e 'l cor per sua magione").[16] Accordingly, each of the three spirits expresses itself in its own style: tragic for the vital spirit, comic for the animal spirit, and elegiac for the natural spirit.[17]

Instead of beginning with the effects exerted by the sight of Beatrice on the animal spirit carried through the visual nerves (*li spiriti del viso*), he first mentions the vital spirit, which is said to dwell in the heart and affect the veins.[18] Dante describes how the vital spirit trembles violently and affects the pulse, making it very feeble (*li mènimi polsi*).[19] The reference to the pulse is medically accurate, as arteries were thought to carry the life-bearing vital spirit along with blood throughout the body, and their movement and vigour – usually assessed by physicians through the pulse – could be affected by strong emotions.

Among the *signa ex parte corporis* due to the vital spirit, Dante mentions trembling ("cominciò a tremare sì fortemente; e tremando").[20] This symptom also occurs to him in the "gabbo" episode in Chapter XIV, where the presence of Beatrice is accompanied by trembling:

> No sooner had I reached this decision [*i.e.* to remain with his friend in attendance upon the ladies] than I seemed to feel a strange throbbing which began in the left side of my breast and immediately spread to

> all parts of my body. Then, pretending to act naturally, I leaned for support against a painted surface that extended along the walls of the house and, fearing that people might have become aware of my trembling, I raised my eyes and, looking at the ladies, I saw among them the most gracious Beatrice.[21]

Although not explicitly mentioned, Dante's symptoms can clearly be attributed to an alteration of the vital spirit, affected in its natural seat (the heart) and its route of diffusion in the body (the arteries). Again, as in his first encounter with Beatrice, Dante first describes the effect on the vital spirit (rather than on the animal spirit through the nerves of sight), and, interestingly, writes that it was affected even before he actually saw her. His tremors, due to the impaired function of the vital spirit, also anticipate the presence of Beatrice.[22]

Beatrice's influence on Dante's body is capable of disrupting and subverting perception and feeling.[23] It is her very presence, wondrous and ineffable, rather than her appearance "seen" by Dante, that elicits these effects in the poet. The blessed woman acts on Dante's body in a miraculous and almost supernatural way (*praeter naturam*), making his life new and wondrous, a true "Vita nuova", even in terms of bodily functioning.[24]

4 Animal Spirit

In describing the seat of the animal spirit, Dante mentions "the high chamber to which all the senses bring their perceptions" ("l'alta camera ne la quale tutti li spiriti sensitivi portano le loro percezioni"), explicitly referring to the medieval cell theory of the inner senses.

The choice of the singular (*camera*, chamber) instead of the plural (*camere*, chambers) could reflect the rhetorical device of synecdoche, which consists of using a term for a part of something to refer to the whole (*pars pro toto*). Similarly, the adjective used to refer to it (*alta*, high) could indicate that it was the seat of the highest power of the soul, the one responsible for reasoning or intelligence. With this regard, in *Convivio,* Dante wrote:

> For, as the Philosopher [i.e., Aristotle] says in the second book of On the Soul, the powers of the soul stand above each other as the figure of the quadrangle stands above that of the triangle, and the pentagon (that is, a figure with five sides) stands above the quadrangle: so the sensitive power stands above the vegetative power, and the intellectual power stands above the sensitive power.[25]

However, it is also possible that in the quoted passage from the *Vita Nuova* Dante referred to a specific cerebral ventricle, the anterior one, which anatomically was located above the others (thus justifying the use of the term *alta camera*, high chamber) and received information from all the sensory nerves that were connected to it. This cavity was responsible for the initial processing of the information conveyed through the nerves and the discrimination of the different modalities of perception (*sensus communis*, common sense). The perception was then transmitted to the posterior part of the anterior ventricle, where images were stored (*imaginatio*, imagination). Common sense (receiving power) and image storage (retaining power) constituted the fantasy (*fantasia*).[26] The anterior ventricle acted as a sensory centre, receiving and processing information from the nerves and then transmitting it to the other cognitive faculties located in the middle ventricle (*vis cogitativa*, cogitation, and *vis aestimativa*, estimation or instinct) and the posterior ventricle (*memoria*, memory, or recollection).[27]

Identifying with precision the sources of Dante's knowledge of medical theories is challenging. However, it is worth noting that the cell doctrine is explicitly mentioned in *Il Tesoretto*, the didactic-allegorical poem written in vernacular by Brunetto Latini, whom Dante recognised as his teacher:

> Nel capo son tre çelle
> e io dirò di quelle.
> Davanti è lo ricetto
> di tutto lo 'ntelletto
> e la forza d'aprendere
> quello che puoi intendere;
> in mezzo è la ragione
> e la discrezïone,
> che cerne ben da male,
> e lo torto e l'iguale;
>
> Di dietro sta con gloria
> la valente memoria,
> che ricorda e ritiene
> quello che 'n esso avene.
> (Brunetto Latini, *Il Tesoretto*, vv. 749–762)

> [In the head are three cells; / I will tell you of these. / In front is the receiver / Of all intelligence / And the ability of understand / That which you can learn; / In the midst is Reason, / And Discretion, / Which discerns good from ill, / The wrong from the just /On the right with glory / Is worthy Memory, / which retains and records /That which happens to it].[28]

It is possible that Dante gained his knowledge of the inner senses from this text, although Brunetto Latini does not refer to the first cell as the sensory centre. Another possible source could be Bartholomaeus Anglicus' *De proprietatibus rerum:*

> There are in fact three cells in the brain: namely, the anterior, in which the imaginative power operates, which organizes and composes

> internally what the sense apprehends from outside, as Johannitius says [i.e., Ḥunayn ibn Isḥāq, a Nestorian Syrian physician and translator of numerous medical treatises, including works attributed to Galen]; there is also a middle cell, that is, logistic, in which sensible or estimative reason dominates; and there is the third and last, which is that of memory, holding and preserving in the treasure of memory the things that have been apprehended by imagination or reason.[29]

Saint Augustine as direct (from *De genesi ad litteram*) or mediated knowledge (through the *Liber de spiritu et anima*) could have been a further source:

> therefore, there are three ventricles in the brain: one anterior toward the face, from which all sense comes; the other posterior to the neck, from which all movement comes; and a third between the two, in which memory is shown to be active.[30]
>
> There are three ventricles in the brain. One anterior, from which all sense [originates]; the other posterior, from which all movement [originates]; the third is the middle between the two, that is, the rational. [...] In the first part of the brain, the animal power is called fantastic (*phantasticum*), that is, imaginary, because in it are contained similitudes and images of corporeal things, whence it is called fantastic. In the middle part of the brain it is called rational, because it examines and judges things that are represented by the imagination. In the last part, it is called memorialis, because there it commends memories that have been judged by reason.[31]

It is therefore not surprising that in the quoted passage, the animal spirit "speaks directly to the spirits of sight". The acts of seeing Beatrice and recalling and visualising her mental image stored within the poet are deeply intertwined and recur frequently in the *Vita Nuova*.

In the Middle Ages, it was unanimously agreed that the animal spirit transmitted visual images to the cerebral ventricles through the hollow visual nerves, but the exact mechanisms of vision were the subject of heated debate, polarised around two opposing medical and philosophical theories.[32] According to the extramissive doctrine, vision occurred through the emission of rays from the eyes that combined with external light to hit an object; sparks were then transmitted back to the eyes, ensuring vision. Conversely, the mediumistic-intramissive theory held that the eye passively received the forms, or "species", of external objects, which had to pass through a transparent or diaphanous medium to become visible.[33]

As is well known, in *Conv.* III, 9, Dante recalls and rejects the extramissive theory of vision, advocating the mediumistic-intramissive doctrine. However, his attitude towards optics in the *Vita Nuova* appears more nuanced and perhaps even ambivalent.

In recounting the episode that will lead to his decision to conceal his love for Beatrice by means of the *donna schermo* (*Vita Nuova* V), Dante employs terms (*mezzo*, medium; *retta linea*; straight line) that are reminiscent of those used in contemporary optics.[34] The use of these terms might suggest that the *donna schermo* represented the metaphorical counterpart of the physical medium through which Dante's vision of Beatrice could be realised respecting the conventions of courtly love.

However, in the *Vita Nuova*, there are possible hints of the extramissive theory of vision. Consider the following passage:

> And when she was about to greet me, one of Love's spirits, annihilating all the others of the senses, would drive out the feeble spirits of sight, saying to them, "Go and pay homage to your mistress", and Love would take their place.[35]

Love's spirits driving out the spirits of sight, commanding them to go to their mistress, is reminiscent of Hugh of St. Victor's account of perception, according to which the soul somehow "rushes out" to grasp external objects.[36] This is also entirely in line with the Galenic version of the extramissive theory, which holds that eyes emit animal spirits that have descended through the optic nerves from the ventricles (hence the emphasis on the connection between the animal spirit residing in the "high chamber" of the brain and the spirits of sight in *Vita Nuova* II).[37]

In the *Vita Nuova*, a further passage that draws on the extramissive doctrine of vision is found in the *canzone* XIX:

> De li occhi suoi, come ch'ella li mova,
> escono spirti d'amore infiammati,
> che feron li occhi a qual che allor la guati,
> e passan sì che 'l cor ciascun retrova.
> (Vita Nuova XIX)

> [her eyes, wherever she may choose to look, / send forth their spirits radiant with love / to strike the eyes of anyone they meet, / and penetrate until they find the heart].

Other references to the extramissive doctrine of vision are found in the *Rime*.[38]

It is perfectly reasonable and justified to state that

> Dante might draw upon theories of extramission when it suits his purposes in poetry, and how he does so in line with vernacular poetic and other authoritative precedents for the fiery eye beam in philosophical and theological works.[39]

However, I think we could also consider the possibility that, at least initially, Dante followed the extramissive doctrine of vision, switching to the mediumistic-intramissive theory at a later stage (surely at the time he wrote the *Convivio*).

5 Natural Spirit

Among the sudden and synchronous effects of Beatrice's first appearance described by Dante, the spirit that speaks last is the natural spirit. It seems to be the spirit that suffers most, as it weeps, complains, and addresses itself as miserable (*Heu miser*), lamenting that henceforth its function will be impaired.

In Chapter IV, Dante further describes the bodily effects of the disrupted function of the natural spirit:

> After that vision my natural spirit was interfered with in its functioning, because my soul had become wholly absorbed in thinking about this most gracious lady; and in a short time I became so weak and frail that many of my friends were worried about the way I looked.[40]

He recalls how the impairment of his natural spirit (probably through inadequate food absorption resulting in decreased blood production by the liver) led to a dramatic change in his appearance. Even his own face showed the destructive signs of Love, provoking the curiosity of others:

> I said that it was Love because there were so many of his signs clearly marked on my face that they were impossible to conceal. And when people would ask: 'Who is the person for whom you are so destroyed by Love?'[41]

In the subsequent chapter (*Vita Nuova* V), those who believe that the *donna schermo* is the cause of Dante's suffering comment: "See what a devastating effect that lady has had on that man" ("Vedi come cotale donna distrugge la persona di costui"). Here, it is Dante's whole body that is devastated and destroyed by lovesickness.[42] Although it is not clearly reported, its bodily effect is probably a marked loss of weight rather than a paling of the skin, a *signum amoris* that, elsewhere in the *Vita Nuova*, Dante very accurately reports in association with an impairment of the vital spirit running through the arteries. According to the medical theories of the time, the disrupted function of the natural spirit had led the poet to a profound state of *maceratio*, which – if continued – could have resulted in physical death.[43]

6 Conclusion: What Is Love?

The *Vita Nuova* revolves around a straight, although elusive, question: what is Love? Dante explicitly addresses this question several times in this work, but, surprising as it may seem, without providing an entirely coherent answer.

At the beginning of Chapter XXV, he states:

> At this point it may be that someone worthy of having every doubt cleared up could be puzzled at my speaking of Love as if it were a thing in itself, as if it were not only an intellectual substance, but also a bodily substance. This is patently false, for Love does not exist in itself as a substance, but is an accident in a substance.[44]

Here Dante refers to Love as an accident, an attribute that is neither necessary nor self-sufficient, as it depends on another substance to subsist.[45] This conception of Love as an accident that qualifies or inheres in a substance was also shared by Guido Cavalcanti.[46] The definition of Love as an accident is justified by the fact that it is not necessarily present in every human being.[47] In this regard, it is noteworthy that, according to medieval medical beliefs, emotions were regarded as accidents of the soul (*accidencia anime*), and the very term accidents (*accidencia*) was used by physicians to refer to concomitant phenomena or symptoms that accompanied diseases without being necessarily present.[48] By defining Love as "an accident in a substance", Dante emphasises that emotions are accidents of the soul substantiated in a body (embodied and enmattered).

Dante further deals with the nature of Love in *Vita Nuova* XX. In the opening sonnet, he states that "Love and the gracious heart are a single thing; [...] one can no more be without the other than can the reasoning mind without its reason". This concept is further expanded in the sonnet and related glosses of Chapter XXI. Dante clarifies the nature of Love relying on the philosophical concepts of action and potentiality. The beloved woman not only actuates/activates/awakens the potentiality of Love that resides in the heart of the loving man but is even able to create such potentiality where none exists. This "miracle too rich and strange" (*novo miracolo e gentile*) draws a parallel between woman and God, conceived as pure Act (*actus purus*), pure form, unmover mover (*primum movens*), capable of creating matter out of nothing (*creation ex nihilo*), infusing form and spirit of life into amorphous and inanimate matter. Within this wondrous conception, the potentiality of love and the activating power of the woman are related to each other "as matter is to form".[49] The intimate and indissoluble unity of matter and form that Dante refers to in this passage is nothing other than the Aristotelian concept of concrete individual substance, defined as a "complex whole" (σύνολον) composed of its matter and form. Love is thus a potentiality activated by the beloved woman in the gracious heart (seat of the vital spirit) of the lover. It ceases to be a mere accident and is miraculously transformed into a substance in which the beloved lady and the lover become inextricably united. The miracle performed by the woman is therefore both a transubstantiation and a transfiguration.[50] Through the creative power of Love (and poetry), the potentiality is activated, the gracious heart becomes Love, matter becomes form, the accident becomes substance, and the human being goes "beyond the sphere that makes the widest round" ("Oltre la spera che più larga gira"), entering the mysterious and ineffable domain of God.

Notes

1 I dedicate this work to the women who renewed my life with the wondrous mystery of beauty and Love.

2 An equally emblematic and well-known example of this reciprocal interaction is the commentary on Guido Cavalcanti's canzone *Donna me prega* made by the physician Dino del Garbo as if it were a medical dissertation. Dino del Garbo, "Glossa alla canzone d'amore", in Fenzi, *La canzone d'amore*, 86–133. Consider also the sharp response, full of medical references, by Dante da Maiano (*Di ciò che stato sei dimandatore*) to Dante's sonnet *A ciascun'alma presa e gentil core* (*Vita Nuova*, III).

3 Among them: Nardi, "L'amore e i medici medievali"; Ciavolella, *La "malattia d'amore"*; Tonelli, *Fisiologia della passione*; Fenzi, *La canzone d'amore*.

4 Whether Dante, while in exile in Bologna, attended the lectures of the Florentine physician Taddeo Alderotti, the *Taddeo ipocratista* (Thaddeus the Hippocratist) quoted in Dante, *Convivio* I, x, 10, is still a matter of debate. See Tonelli, "Medicine".

5 In a scientific and truthful manner, drawing on the precepts of natural science. Dino del Garbo, "Glossa alla canzone d'amore", in Fenzi, *La canzone d'amore,* 88–89.

6 This distinction was adopted by Gerard of Berry in his commentary on the *Viaticum peregrinantis* written around 1180–1200. See: Gerard de Berry, *Glosule super Viaticum*. The *Viaticum* was the Latin translation of the Arabic medical text written by Ibn al-Djazzar, originally entitled *Zād al-musāfir*. This work, conceived as a manual for travelers who did not have access to medical care, was translated by Constantine the African and enjoyed great popularity.

7 *Vita Nuova,* II: "In quello punto dico veracemente che lo spirito de la vita, lo quale dimora ne la secretissima camera de lo cuore, cominciò a tremare sì fortemente, che apparia ne li menimi polsi orribilmente; e tremando disse queste parole: *Ecce deus fortior me, qui veniens dominabitur michi.* In quello punto lo spirito animale, lo quale dimora ne l'alta camera ne la quale tutti li spiriti sensitivi portano le loro percezioni, si cominciò a maravigliare molto, e parlando spezialmente a li spiriti del viso, sì disse queste parole: *Apparuit iam beatitudo vestra.* In quello punto lo spirito naturale, lo quale dimora in quella parte ove si ministra lo nutrimento nostro, cominciò a piangere, e piangendo disse queste parole: *Heu miser, quia frequenter impeditus ero deinceps!*". All texts from *Vita Nuova* reported in this chapter are taken from the edition by Michele Barbi. The English translation used in this chapter is that by Mark L. Musa.

8 Surprisingly, in the entire corpus of Dante's works, explicit references to humoural theory are extremely scarce. In the *Vita Nuova* only one passage mentions it, that is, *Vita Nuova*, XXXI: "No quality of heat or cold took her away from us [i.e., Beatrice], as is the fate of others [*i.e.*, other women]" ("no la ci tolse qualità di gelo / né di calore, come l'altre face"). Indirect allusions to the medical theory of the four humours and in particular to the melancholic complexion have been extensively studied by Tonelli, "Medicine", 227–241 and Tonelli, *Fisiologia della passione,* 135–144.

9 It has been suggested that the Galenic theory of the three spirits gained more popularity after the First Council of Nicaea (325 AD), which confirmed the trinitarian dogma and condemned the doctrine of Arius that upheld the supremacy and uniqueness of God the Father and the subordination of the Son. See Cosmacini,

Dante e l'arte medica, 14. In describing the marvellous effects caused by Beatrice's almost supernatural appearance, Dante might have drawn a parallel with the divine trinity.

10 A detailed description of this medical doctrine is provided in Bartholomaeus Anglicus, *De proprietatibus rerum*, III, 12; III, 14. This encyclopaedic work (dated about 1240), intended for the use of students and the general public, was widely quoted in the Middle Ages.

11 Quotations from the *Commedia* reported in this chapter are taken from the edition by Petrocchi. English translations are by Singleton.

12 Actually, these cavities are filled with liquid. The ancient belief of hollow brain cavities was the consequence of faulty dissection techniques: removal of the head from the trunk emptied the brain of the fluids contained in the ventricles. It was only in 1764, thanks to refinement of dissection techniques by Domenico Cotugno (1736–1822), that the presence of fluid in the cerebral ventricles was definitively established.

13 An in-depth account of this theory, from its inception to its subsequent development, can be found in Kemp and Fletcher, "The Medieval Theory of the Inner Senses". See also: Wolfson, "The Internal Senses"; Kemp, *Cognitive Psychology* and Smith, Frixione, Finger, and Clover, *The Animal Spirit Doctrine*, 71–78. A series of extraordinary drawings illustrating the medieval cell doctrine can be found in Clarke and Dewhurst, *An Illustrated History of Brain Function.*

14 *Vita Nuova*, XXV: "Dunque, se noi vedemo che li poete hanno parlato a le cose inanimate, sì come se avessero senso e ragione, e fattele parlare insieme; e non solamente cose vere, ma cose non vere, cioè che detto hanno, di cose le quali non sono, che parlano, e detto che molti accidenti parlano, sì come se fossero sustanzie e uomini; degno è lo dicitore per rima di fare lo somigliante, ma non sanza ragione alcuna, ma con ragione la quale poi sia possibile d'aprire per prosa". Incidentally, although Tonelli (*Fisiologia della passione*, 143) states that "the case of inanimate beings conversing with one another is never contemplated in the poetic texts of the book", this type of personification certainly applies to the prose text of *Vita Nuova* II.

15 *Ecce deus fortior me, qui veniens dominabitur michi* ("Here is a god stronger than I who comes to rule over me"). In Matthew's Gospel (Mt 3,11) while baptising, John the Baptist said: *qui autem post me venturus est, fortior me est* ("but he that cometh after me is mightier than I").

16 *Vita Nuova* XX: "Love and the gracious heart are a single thing". Many other examples from Dante's works could be provided. Among the most famous, is *Inf.* V, 100: "Amor, ch'al cor gentil ratto s'apprende" ("Love, which is swiftly kindled in the noble heart").

17 Rossi, "Commento" a *Vita Nova*, 9–10. See also: Dante, *De vulgari eloquentia* II, 4.5–6.

18 In this passage, Dante should have referred to arteries rather than veins. However, as in the *Commedia*, Dante uses the term veins even when clearly referring to arteries. The term *arteria* (artery) never appears in the entire corpus of Dante's works. According to *TLIO, Tesoro della Lingua Italiana delle Origini* (online), the term *artere* (artery) is attested only once before the *Vita Nuova*. See: *TLIO, Tesoro della Lingua Italiana delle Origini,* 18 July 20263, http://tlio.ovi.cnr.it/TLIO/

19 The term *polsi* (pulses), from the verb *pellere*, to push, expel, set in motion, refers to the effect of blood mixed with the vital spirit circulating in the arteries. Sometimes Dante uses this term to refer to the arteries, as in *Inferno* I, 90: "ch'ella mi fa tremar le vene e i polsi" ("for she makes my veins and pulses tremble").

20 Heart tremor and pulse disturbances were often associated. See: Bartholomaeus Anglicus, *De proprietatibus rerum* III, 24: "fear likewise makes the pulse fast and irregular and trembling" ("timor similiter pulsum efficit velocem et inordinatum et trementem"). Dante refers to this trembling of the heart also in other parts of

the *Vita Nuova*. In chapter XVI: "nel cor mi si comincia uno tremoto, / che fa de' polsi l'anima partire" ("within my heart a tremor starts to spread, driving out life, stopping my pulses' beat"); and in chapter XXIV: "come io mi sentì svegliare lo tremore usato nel cuore" (how I felt the familiar tremor awaken in my heart). See also Tonelli, *Fisiologia della passione,* 74–77 and 82–83.

21 *Vita Nuova*, XIV: "E nel fine del mio proponimento mi parve sentire uno mirabile tremore incominciare nel mio petto da la sinistra parte e distendersi di subito per tutte le parti del mio corpo. Allora dico che io poggiai la mia persona simulatamente ad una pintura la quale circundava questa magione; e temendo non altri si fosse accorto del mio tremare, levai li occhi, e mirando le donne, vidi tra loro la gentilissima Beatrice".

22 The physical reaction of trembling will be evoked in the encounter with Beatrice in *Purg*. XXX, 46–48: "Men che drama / di sangue m'è rimaso che non tremi: / conosco i segni de l'antica fiamma" ("Less than a drop of blood is left me that does not trembe: I recognize the signs of the ancient flame").

23 Interestingly, the sequence of bodily reactions described by Dante in this episode is strikingly similar to that outlined by the James-Lange theory, proposed in the 19th Century by William James (1842–1910) and Carl Lange (1834–1900). According to this theory, bodily changes can precede and induce the affective experience of emotion. Remarkably, Thomas Aquinas conceived the soul as the subject and cause of passions. Thomas Aquinas, *Summa Theologiae* I^{a}-IIae q. 37 a. 4 ad 1: "since the soul naturally moves the body, the spiritual movement of the soul is naturally the cause of bodily transformation" ("quia anima naturaliter movet corpus, spiritualis motus animae naturaliter est causa transmutationis corporalis"). In the "gabbo" episode, Dante seems to reverse this order, making his bodily reaction to Beatrice's presence the cause of its subsequent emotional experience. One final suggestion: etymologically, the term emotion (Latin: *emotio*) is derived from the verb *emovere*, to move out. Another clue to support the intimate relationship between emotions and their behavioural correlates? Why not. As Dante would have said, *nomina sunt consequentia rerum* (*Vita Nuova*, XIII).

24 I intentionally avoided to use the term "physiology" because it had a completely different meaning in the Middle Ages than it has nowadays.

25 See *Conv.* IV, vii, 14: "Ché, sì come dice lo Filosofo nel secondo de l'Anima, le potenze de l'anima stanno sopra sé come la figura de lo quadrangulo sta sopra lo triangulo, e lo pentangulo, cioè la figura che ha cinque canti, sta sopra lo quadrangulo: e così la sensitiva sta sopra la vegetativa, e la intellettiva sta sopra la sensitiva". English transl. by Hillard.

26 The disruptive effects of Love on Dante's *imaginatio* and *fantasia* in the *Vita Nuova*, and its association with dreams and visions of prophetic value, have been thoroughly investigated by Tonelli, "Medicine", 227–241 and Tonelli, *Fisiologia della passione*, 135–144.

27 Kemp, *Cognitive Psychology*, 45–63.

28 Brunetto Latini, *The Little Treasure*, 40–41.

29 Bartholomaeus Anglicus, *De proprietatibus rerum* III, 10: "Nam in cerebro sun tres cellule: scilicet anterior, in qua virtus imaginativa operantus, que quidem ea que sensus extrinsecus apprehendit interius ordinat et componit, ut dicit Iohannitius; est et media cellula, scilicet logistica, in qua ratio sensibilis sive estimativa virtus dominatur; est iterum et tertia et postrema, que est memorativa, que ea que apprehensa sunt per imaginationem sive rationem in thesauro memorie retinet et custodit". See also: Bartholomaeus Anglicus, *De proprietatibus rerum* III, 16 and 22.

30 Augustine, *De Genesi ad litteram*, VII, 18–24: "ideo tres tamquam ventriculi cerebri demonstrantur: unus anterior ad faciem, a quo sensus omnis; alter posterior ad cervicem, a quo motus omnis; tertius inter utrumque, in quo memoriam vigere demonstrant".

31 Augustine [pseudo], *Liber de Spiritu et Anima*, XII, 796: "Tres namque sunt ventriculi cerebri. Unus anterior, a quo omnis sensus; alter posterior, a quo omnis motus; tertius inter utrumque medius, id est, rationalis. [...] In prima parte cerebri vis animalis vocatur phantastica, id est imaginaria; quia in ea corporalium rerum similitudines et imagines continentur, unde et phantasticum dicitur. In media parte cerebri vocatur rationalis; quia ibi examinat et judicat ea quae per imaginationem representantur. In ultima parte vocatur memorialis; quia ibi commendat memoriae quae a ratione sunt judicata". The *Liber de Spiritu et Anima* is a collection of texts (*cento*) from different authors (including Saint Augustine, Boethius, Isidore of Seville, Bede, Bernard of Clairvaux, and Hugh of St. Victor) that had a wide circulation in the Middle Ages as a devotional text. In my opinion, further research is needed to clarify its possible echoes in Dante's works.

32 During the Middle Ages and up to the threshold of the modern age, the distinction between medicine and natural philosophy (*philosophia naturalis*) was blurred. The term *physicus* (from the Greek φύσις, nature) was used as a synonym for *medicus* since the early 12th Century. See: Tonelli, *Fisiologia della passione*, 8.

33 See Lindberg, *Theories of Vision*; Gilson, "Visual Theory"; Tonelli, *Fisiologia della passione*, 31–36; Kemp, *Medieval Psychology.*

34 Ardizzone, "Optics".

35 *Vita Nuova* XI: "E quando ella fosse alquanto propinqua al salutare, uno spirito d'amore, distruggendo tutti li altri spiriti sensitivi, pingea fuori li deboletti spiriti del viso, e dicea loro: 'Andate a onorare la donna vostra'; ed elli si rimanea nel luogo loro. E chi avesse voluto conoscere Amore, fare lo potea mirando lo tremare de li occhi miei".

36 Hugh of St. Victor, *Didascalicon* II, 3: "Through contact with physical objects it degenerates, because, while through sense impressions it rushes out toward the visible forms of bodies and, having made contact with them, draws them into itself through imagination, it is cut away from its simplicity each time it is penetrated by any qualities entering through hostile sense experience" ("Tangendo ergo corpora degenerat, quia, dum in visibiles corporum formas per sensuum passiones procurrit easque attactas per imaginationem in se trahit, toties a sua simplicitate scinditur, quoties aliquibus contrariae passionis qualitatibus informatur. Cum vero ab hac distractione ad puram intelligentiam conscendens in unum se colligit, fit beatior intellectibilis substantiae participatione").

37 Galen, *On the Doctrines of Hippocrates and Plato* VII, 4.4–11, 448–451.

38 Dante, *Rime,* LXV: "Degli occhi della mia donna si move / un lume sì gentil che dove appare" ("A light moves from my lady's eyes so noble that where it is appears"); and XCI: "Entrano i raggi di questi occhi belli / ne' miei innamorati" ("The rays of these beautiful eyes enter into my loving eyes").

39 Gilson, "Visual Theory", 248–249.

40 *Vita Nuova*, IV: "Da questa visione innanzi cominciò lo mio spirito naturale ad essere impedito ne la sua operazione, però che l'anima era tutta data nel pensare di questa gentilissima; onde io divenni in picciolo tempo poi di sì fraile e debole condizione, che a molti amici pesava de la mia vista; e molti pieni d'invidia già si procacciavano di sapere di me quello che io volea del tutto celare ad altrui".

41 *Vita Nuova*, IV: "Dicea d'Amore, però che io portava nel viso tante de le sue insegne, che questo non si potea ricovrire. E quando mi domandavano 'Per cui t'ha così distrutto questo Amore?', ed io sorridendo li guardava, e nulla dicea loro".

42 On Love as a sickness (including its variant known as *ereos*, corresponding to what Avicenna called *ilisci*) see: Tonelli, *Fisiologia della passione,* 15–25. See also: Ciavolella, *La "malattia d'amore".*

43 See: Arnaldus de Villanova, *Tractatus de amore heroico*, 51–52: "Therefore, for these reasons, the limbs of the body become emaciated, especially those that are rarer and softer and those that are more subject to resolution, whereby the face becomes thinner and the eyes hollow" ("Macerantur igitur hiis de causis corporis

membra, precipue rariora et molliora et ea que magis sunt resolutioni subiecta, quamobrem facies extenuatur et oculi concavantur"). See also: Tonelli, *Fisiologia della passione,* 79–81; Nardi, "L'amore e i medici medievali", 517–542; Ciavolella, *La "malattia d'amore"*; Lowest, "The Loveres Maladye"; Lacarra Lanz, "El 'amor que dicen hereos'".

44 *Vita Nuova*, XXV: "Potrebbe qui dubitare persona degna da dichiararle onne dubitazione, e dubitare potrebbe di ciò, che io dico d'Amore come se fosse una cosa per sé, e non solamente sustanzia intelligente, ma sì come fosse sustanzia corporale: la quale cosa, secondo la veritate, è falsa; ché Amore non è per sé sì come sustanzia, ma è uno accidente in sustanzia".

45 See Thomas Aquinas, *Summa Theologiae* I[a] q. 28 a. 2 co.: "accidentis proprium est inesse".

46 See Guido Cavalcanti's canzone *Donna me prega,* vv. 1–3, in Fenzi, *La canzone d'amore*: "A lady bids me, and so I want to speak of an accident that is often fierce and is so noble that [it] is called 'love' [amore]" ("Donna me prega, - per ch'eo voglio dire / d'un accidente - che sovente - è fero / ed è sì altero - ch'è chiamato amore").

47 See Dino del Garbo's "Glossa alla canzone d'amore" in Fenzi, *La canzone d'amore,* 90–91: "Now this passion is called an accident, first, because it is not a substance in itself, but – being an appetite of the soul – is attached to another as its subject, in the same way as the passions of the soul, which are anger, sadness, fear, and the like. Second, it is called an accident because it can come and go away like other accidents. Third, it is called an accident, because it comes from outside, and although someone may have an intrinsic disposition by which he easily incurs this suffering, as will be explained later, the cause itself is first and foremost an external thing" ("Dicitur autem hec passio accidens, primo quia non est substantia per se stans, sed est alteri adherens sicut subiecto, ut appetitus anime, simili modo sicut anime passiones, que sunt ira, tristitia, timor et similis. Secundo dicitur accidens, quia potest advenire et etiam recedere sicut accidentia alia. Tertio dicitur accidens, quia advenit ab aextrinseco et, licet secundum aliquid possit quis habere dispositionem intrinsecam per quam faciliter incurrat in hanc passionem, ut postea declarabitur, tamen causans ipsum principaliter est res extrinseca").

48 Demaitre, *Medieval Medicine.* See also: Peter of Spain, *Questiones super Viaticum*, 216: "Similarly, all the accidents of the soul are the passion of the heart, since all the accidents of the soul are the result of the heart, as Haly says above Tegni. But the love *hereos* is an accident. Therefore, it is a passion of the heart" ("Item omnia accidencia anime sunt passio cordis, quoniam omnia accidencia anime consequuntur cor ut dicit Haly super Tegni. Sed amor hereos est accidens. Ergo est passio cordis").

49 *Vita Nuova*, XX: "l'uno guarda l'altro come forma materia".

50 It is a transubstantiation, as the substance of the lover, his heart – which stands as a synecdoche to represent his entire emotional life – is transformed into something new ("Vita nuova") by the beloved lady, who activates the potentiality of Love that dwells in him. It is also a transfiguration (*trasfigurazione*, *trasfiguramento*: *Vita Nuova*, XIV), as the lover's *accidencia anime*, which manifest as *signa amoris ex parte corporis et ex parte animae,* are affected and transformed (*trans*, across, and *forma*, form, aspect) by the beloved woman.

Bibliography

Primary Sources

Arnaldus de Villanova. *Tractatus de amore heroico*. Edited by Michael R. McVaugh. Barcelona: Publicacions i edicions de la Universitat de Barcelona, 1985.

Augustine [pseudo]. *Liber de Spiritu et Anima*. Edited by Jacques Paul Migne. *Patrologia Latina*, vol. 40, coll. 779–832. Paris: I. P. Migne, 1845.

Augustine. *De Genesi ad litteram*. Edited by Joseph Zycha. Wien: Tempsky, 1894.

Bartholomaeus Anglicus. *De proprietatibus rerum*. Edited by Christel Meier, Heinz Meyer, Baudouin Van Den Abeele, and Iolanda Ventura, vol. 1. Turnhout: Brepols, 2007.

Brunetto Latini. *Il Tesoretto* (*The Little Treasure*). Translated by Julia Bolton Holloway. New York and London: Garland Publishing, Inc., 1981.

Brunetto Latini. *Il Tesoretto*. Edited by Marcello Ciccuto. Milan: Rizzoli, 1985.

Dante Alighieri. *Convivio (The Banquet)*. Translated by Katharine Hillard. London: Kegan Paul, Trench & Co., 1889.

Dante Alighieri. *Convivio*. Edited by Antonio Enzo Quaglio. Florence: Felice Le Monnier, 1964.

Dante Alighieri. *Inferno*. Edited by Giorgio Petrocchi. Dante. *La comedia secondo l'antica vulgata,* vol. 2. Florence: Le Lettere, 1994.

Dante Alighieri. *Rime*. Edited by Michele Barbi. *Le Opere di Dante*. Florence: Società Dantesca Italiana, 1960.

Dante Alighieri. *The Divine Comedy. Inferno*. Translated, with a commentary, by Charles S. Singleton. London: Routledge & Kegan Paul, 1971.

Dante Alighieri. *Vita Nuova*. Edited by Michele Barbi. *Le Opere di Dante*. Florence: Società Dantesca Italiana, 1960.

Dante Alighieri. *Vita Nuova*. Translated by Mark L. Musa. Bloomington: Indiana University Press, 1973.

Galen. *On the Doctrines of Hippocrates and Plato. Second Part. Books VI–IX*. Edited by Philipp de Lacy. Berlin: Akademie Verlag GmbH, 2005.

Gerard de Berry. *Glosule super Viaticum*. Edited by Mary F. Wack. *Lovesickness in the Middle Ages. The Viaticum and Its Commentaries*, 194–205. Philadelphia: University of Pennsylvania Press, 1990.

Hugh of St. Victor. *Didascalicon*. Translation with introduction and notes by Jerome Taylor. New York and London: Columbia University Press, 1961.

Peter of Spain. *Questiones super Viaticum*. Edited by Mary F. Wack. *Lovesickness in the Middle Ages. The Viaticum and Its Commentaries*, 83–108. Philadelphia: University of Pennsylvania Press, 1990.

Thomas Aquinas. *Summa Theologiae [cura fratrum eiusdem Ordinis]*. Madrid: La Editorial Catolica, 1961.

Secondary Sources

Ardizzone, Maria Luisa. "Optics". In *The Dante Encyclopedia*, edited by Richard Lansing, 660–662. London and New York: Routledge, 2010.

Ciavolella, Massimo. *La "malattia d'amore" dall'Antichità al Medioevo*. Rome: Bulzoni, 1976.

Clarke, Edwin and Kenneth Dewhurst. *An Illustrated History of Brain Function*. Oxford: Sanford Publications, 1972.

Cosmacini, Giorgio. *Dante e l'arte medica*. Milan: Pantarei, 2021.

Demaitre, Luke. *Medieval Medicine. The Art of Healing, from Head to Toe*. Santa Barbara, CA, Denver, CO and Oxford: Praeger, 2013.

Fenzi, Enrico. *La canzone d'amore di Guido Cavalcanti e i suoi antichi commenti*. Genoa: il Melangolo, 1999.

Gilson, Simon. "Visual Theory". In *The Oxford Handbook of Dante*, edited by Manuele Gragnolati, Elena Lombardi, and Francesca Southerden, 242–256. Oxford: Oxford University Press, 2021.

Kemp, Simon and Garth J. O. Fletcher. "The Medieval Theory of the Inner Senses". *The American Journal of Psychology* 106/4 (1993): 559–576.

Kemp, Simon. *Medieval Psychology.* Westport, CT: Greenwood Press Inc., 1990.

Kemp, Simon. *Cognitive Psychology in the Middle Ages.* London: Greenwood Press, 1996.

Lacarra Lanz, Eukene. "El 'amor que dicen hereos' o *aegritudo amoris*". *Cahiers d'études hispaniques médiévales* 1/38 (2015): 29–44.

Lindberg, David C. *Theories of Vision from al-Kindi to Kepler.* Chicago, IL and London: Chicago University Press, 1976.

Lowest, John Livingston. "The Loveres Maladye of Hereos". *Modern philology* 11 (1913–1914): 491–546.

Nardi, Bruno. "L'amore e i medici medievali". In *Studi in onore di Angelo Monteverdi*, vol. 2, 517–542. Modena: Società tipografica editrice modenese, 1959.

Rossi, Luca Carlo. "Commento" to Dante Alighieri. *Vita Nova*, edited by Luca Carlo Rossi, 1–207. Milan: Mondadori, 2016.

Smith, C. U., Eugenio Frixione, Stanley Finger, and William Clover. *The Animal Spirit Doctrine and the Origins of Neurophysiology.* New York: Oxford University Press, 2012.

Tonelli, Natascia. *Fisiologia della passione. Poesia d'amore e medicina da Cavalcanti a Boccaccio.* Florence: SISMEL-Edizioni del Galluzzo, 2015.

Tonelli, Natascia. "Medicine". In *The Oxford Handbook of Dante*, edited by Manuele Gragnolati, Elena Lombardi, and Francesca Southerden, 227–241. Oxford: Oxford University Press, 2021.

Wolfson, Harry Austryn. "The Internal Senses in Latin, Arabic, and Hebrew Philosophic Texts". *The Harvard Theological Review* 28/2 (1935): 69–133.

9 Dante, Healthcare and Diseases

Michele A. Riva and Lorenzo Lorusso

1 Introduction: Dante's Possible Pathways to Medicine

In his works, Dante Alighieri undoubtedly revealed significant expertise in various fields of medieval science and culture, including medicine.[1] Although the matter is still open to debate, it is possible that starting in his youthful years, he progressively acquired medical and pharmacological knowledge through personal interactions or by reading and attending lectures by scholars in Italian cities where courses on medicine were held, such as Bologna, Siena, and Padua, if not Paris.[2] In particular, Bologna offered the most authoritative instruction in the fields of law and medicine. These disciplines first acquired definite cultural and institutional recognition in Bologna, which led to the creation of the Faculty of Arts and Medicine; the latter was bound to compete for importance and attraction with the long-established Faculty of Laws among students of various European countries. The Italian writer Mario Tobino suggested that Dante arrived in Bologna "with the desire to prey on what was known about the world".[3] It is impossible to ascertain Dante's expectations of the Bolognese intellectual milieu (if any), and it remains a controversial matter whether he stayed in Bologna and attended courses on philosophy, law, and medicine between 1285 and the first months of 1287 (when he was in his twenties).[4] Undoubtedly, Bologna presented a cultural milieu that was "proximate to Dante". In the last decade of the 13th century, when Dante mentioned attending "disputations", certain Italian *philosophantes*, including possibly Jacopo da Pistoia and certainly Gentile da Cingoli, gained early professional recognition in Bologna, particularly in the field of medicine. As Andrea Tabarroni recently underlined, by the end of the 13th century in Bologna and later in Padua, Siena, and Perugia, the distinctive feature of philosophical instruction was indeed its close connection to medicine, which in turn led to a particular focus on methodological issues, the scientific framing of experiences and practice, and the comparison and reconciliation of medicine with the philosophical tradition.[5] Therefore, Dante might have been introduced – if not directly – to lessons in medicine, at least to the rudiments of medical culture, which in his time and specifically in Bologna was closely connected to the teaching of arts and philosophy.

DOI: 10.4324/9781003412601-10

The city itself reflected its cultural environment, enriched by gifts and tributes from political authorities. This was reminiscent of the cultural scene in Paris, where authors translated scientific texts from Greek and Arabic sent by Manfred of Sicily, son of Emperor Frederick II of Swabia.[6] In Bologna, Dante might have followed lectures or disputations by his fellow countryman, the Florentine Taddeo Alderotti. The latter was famous for the logical-dialectical orientation of his teaching, which was stimulated by the medical learning of Hippocrates, Galen, and Aristotle, and aimed to evaluate the autonomy of medical thought based on reason and justified by observation.[7]

Alderotti, who began teaching in Bologna in 1260, is considered one of the most illustrious doctors of his age. He played an important role in the emergence of medicine as a separate and institutionalised discipline that was distinct from albeit connected to philosophy and imbued with high cultural prestige.[8] He wrote *On Preserving Health* (*Della conservazione della salute*), one of the first medical texts in vernacular language. This work followed the precepts of the Salerno School and emphasised the importance of dental hygiene and the promotion of physical activity, particularly among school-aged children. These efforts positioned Alderotti as one of the early pioneers of social medicine. Following the path of the Hippocratic corpus, Alderotti collected approximately 156 clinical cases in his *Consilia*. He was also a commentator of Galen.[9] Dante mentions him in the *Convivio* (I, x, 10) and the *Comedia* (albeit with a somewhat depreciative connotation in the latter case) in the episode of the encounter with St. Dominick. Dominick is said to have endeavoured to become a doctor for the true love of celestial wisdom and not for personal ambition, according to renowned scholars such as "Taddeo", who is generally identified with the Bolognese physician:

> Non per lo mondo, per cui mo s'affanna
> di retro ad Ostiense e a Taddeo,
> ma per amor de la verace manna
> in picciol tempo gran dottor si feo.
> (*Par.* XII, 82–85)

> [Not for this world, for which men toil today, / following Taddeo and the Ostian, /but for love of the true manna / he soon became a teacher so renowned].

Alderotti's best-known pupils were the anatomists Mondino de' Liuzzi and Fiduccio de' Milotti. Mondino taught and practiced in Bologna, where he supervised the first church-approved public dissection session recorded in January 1315. In 1316, the first dissection guide appeared in his work, *Anothomia*.[10] It has been suggested that an acquaintance between Dante and Mondino was possible, though this remains uncertain.[11] However, Dante's friendship with Fiduccio de' Milotti was unquestionable; Fiduccio settled in Ravenna in 1300 and died in Bologna in 1323 while visiting Mondino. Indeed, Fiduccio was Dante's personal physician.[12]

If Dante's familiarisation with medicine can be traced to his early years in Bologna, it is plausible that this milieu influenced his subsequent choice to become a member of the Guild of Physicians and Apothecaries (*Arte dei medici e degli speziali*) once he returned to Florence. Due to his temperament, profile, and relative lack of money or wealth, Dante had two choices when opting for a guild: either among judges and notaries or physicians and pharmacists. Given his education in liberal arts and the basis for philosophy and medicine, he naturally chose the latter. Indeed, Dante is almost always depicted as a physician: with a long, wide red robe (*lucco*); his head covered with a hood (*becchetto*); and the tips falling to the sides of his face.[13] As one of the seven major guilds of Florence, the physicians allowed access to the civil magistracy,[14] a necessary step in participating in Florentine public life.[15] Besides being entrusted with public health and mercantile tasks, the guild was also involved in various cultural and artistic activities. In addition, it also exhibited characteristics of an association catering to the intellectual elites, including poets, philosophers, painters, and writers. These individuals would convene in the hospital apothecary, serving as a gathering place for intellectual exchange and collaboration. Dante seems to have been a frequent participant in apothecaries. Some guild members were prominent public figures who shared Dante's cultural values. Between the 14th and 15th centuries, the guild also admitted established painters such as Giotto, Masaccio, and Paolo Uccello.

Before Dante's exile from Florence, his frequentation of religious institutions in his native town, particularly the convents of Santa Croce and Santa Maria Novella and their rich libraries, likely enriched his culture in natural philosophy and possibly medicine. In this regard, it must be considered that, in Dante's time, medical culture was also largely present in philosophical treatises and commentaries, such as the works of Albert the Great and Thomas Aquinas. Moreover, natural philosophy and its various dimensions had a wide circulation in homilies and sermons held by friars and prelates.[16]

Thanks to these resources, Dante became directly or indirectly acquainted with at least the major medical works of the past. As is well known, the Limbo in the first circle of Hell 'hosts' the four important founders of medicine: the great Muslim doctors Avicenna and Averroes, Hippocrates, and Galen (*Inf.* IV, 143–144: "Ipocràte, Avicenna e Galieno, Averoìs, che 'l gran comento feo"), along with many other ancient scholars and poets, Virgil included, who did not receive baptism (*Inf.* IV, 35: "non ebber battesmo").

The *Canon of Medicine* by Avicenna was translated by Gerard of Cremona in the 12th century. This masterpiece contains the medical principles of Hippocrates, Dioscorides, Galen, and the more recent Alexandrian physicians, along with the biological theories of Aristotle and his collaborators. It exerted a considerable influence on medical faculties, which adopted the treatise as a teaching manual until the emergence of Paracelsus and Vesalius's works in the 16th century. Avicenna was considered the *princeps medicorum*, and his *Canon*, along with the Hippocratic *Aphorisms* and the *Ars parva galenica*, constituted the 'sacred library' for both Eastern and Western medical students. Another of Avicenna's books was the *Canticum de medicina*, which was also translated into Latin by Gerard of Cremona. It is a compendium of medicine written in poetic form to facilitate learning for students of the time. In turn, Averroes wrote the *Colliget* (the "Book of General Principles") on anatomy, health, pathology, symptomatology, dietetics, medicines, hygiene, and therapy. He also penned a summary of Galen's works, a commentary on Avicenna's *Canticum de medicina*, and a treatise on theriac.

Hippocratic medicine proceeded along a scholastic orientation and represented the principle of medical instruction in Italy and the rest of Europe well before Dante's time. In *Par.* XI, Dante mentions the *Aphorisms* as the most representative form of medical writing (vv. 4–5: "Chi dietro a iura, e chi ad amforismi sen giva" / "One pursued the law, one the Aphorisms"), while he explicitly considers Hippocrates as the most accomplished physician in *Purgatory* XXIX, with Saint Luke as one of his followers:

> L'un si mostrava alcun de' famigliari
> di quel sommo Ipocràte che natura
> a li animali fé ch'ell'ha più cari.
> (*Purg.* XXIX, 136–138)

> [One showed himself conjoined with those / who follow great Hippocrates, / whom nature shaped for creatures she loves most].

In *Convivio* I, viii, 5, Dante also mentions Hippocrates' *Aphorisms* and Galen's *Tegni*. In his writing, the Italian poet asserts that it would be inappropriate for a physician to give these books to a knight and for a knight to give a physician a shield ("come quando uno cavaliere donasse ad uno medico uno scudo, e quando uno medico donasse a uno cavaliere inscritti li *Amphorismi* d'Ipocràs o vero li *Tegni* di Galieno"). The same dichotomy between the art of human healing and the "art" of war is also present in the passage from the *Purgatory* quoted above; in the following verse, Dante compares the physician Saint Luke with the 'warrior' Saint Paul, displaying their "opposite care" (v. 139: "mostrava l'altro la contraria cura"), which refers to the sword.

Galen was considered the second highest authority after Hippocrates in medicine, mainly because he had brought the Hippocratic method to fruition within a system enriched by the ideas of Plato and Aristotle, which enabled medicine to have equal authority as philosophy and religion. The *Tegni* mentioned by Dante is a compilation summary written for medical students and was largely in use in medical faculties during Dante's time. Its title is an abridged vernacular form that resounds with the Greek title *Τέχνη ἐχτρική*. The latter was first translated into Latin from Arabic by Constantine the

African with the Latin title *Ars parva* (*Articella* in the vernacular).[17] Galen was also mentioned in the *Monarchia* (I, xiii, 6). However, the effective Galenian text that Dante referred to has been overlooked to date due to an erroneous identification of the source. The reference involves an explicit citation from Galen, the only one in Dante's works. Previous studies have traced it to Galen's *De cognoscendis curandisque animi morbis*, but the idea that Dante could have directly known it is problematic. Surprisingly, Andrea Tabarroni discovered a nearly identical quotation in a celebratory speech written for a graduation ceremony in Bologna in the early 14th century. This discovery consents to identifying the real source in Galen's *De ingenio sanitatis*, book 7.[18] Discoveries such as these underscore how the cultural environment of Bologna may have influenced Dante's immersion in the medical field, even in the absence of a formal academic course of study undertaken by him.

Another celebrated Italian medical school of the time was the aforementioned School of Salerno. This old institution had introduced hygienic and dietary precepts based on the principles of *vis medicatrix naturae* and the doctrine of the four humours, which were both derived from Hippocrates. One of the first surgeons of the School of Salerno was Rogerius of Salerno (died c. 1195), whose *Practice of Surgery* gained popularity by the end of the 12th century. These doctrines were adopted and spread by Italian surgeons, including Rolando Capelluti or Capezzuti from Parma.[19] Dante mentions neither the Salerno School nor any famous surgeon of his time, such as William of Saliceto, in his works. Yet, it is held that he was familiar with the medical works of Peter of Spain (Petrus Hispanus), who is believed to have studied theology and medicine in Paris, also as a pupil of Albert the Great, and in Montpellier and Salerno. In 1247, Peter became a professor at the new School of Medicine at the University of Siena.[20] His many medical writings included the *Thesaurus pauperum,* which was extensively disseminated in manuscripts, prints, and summaries up to the 17th century, and the *Liber de oculo* (or *De oculis*), a treatise on ophthalmology from which Dante may have drawn notions of physiological optics, particularly the visual defect of presbyopia mentioned in *Inf.* XV, 20–21: "They peered at us with knitted brows / like an old tailor at his needle's eye" ("e sì ver' noi aguzzavan le ciglia / come 'l vecchio sartor fa ne la cruna").[21] However, in the *Comedia*, Dante does not mention Peter of Spain's former occupation as a physician, but he does exalt him among the *sapientes* in the Heaven of the Sun for his famous *Summalae logicales*: "Peter the Spaniard / who casts light from his twelve books below" (*Par.* XII, v 134–135: "Pietro Spano / lo qual giù luce in dodici libelli").

2 Healthcare

After the socio-political disintegration of the Roman Empire, the Christian religion established itself through the Church of Rome as a "universal ideal" and a unifying element of the Western world. Through its parishes and monasteries, the Church controlled and administered territories and endeavoured to meet the needs of the population. Particular attention was given to the sick and the poor (who were considered images of God's suffering on earth), and places called "xenodochy" (already known in the Greek and Roman periods) were founded to receive them. The xenodochies were initially connected to churches and consisted of a single room where medical assistance was administered, albeit not yet within a well-defined scientific scope.

The idea of a hospital that was intended as a structure for the recovery of the sick had not yet emerged in early medieval society, and curing disease was not pursued with any real methodology. It was the monastic orders that dedicated themselves to the construction of the first hospitals. These were usually set within gardens, where medical herbs were cultivated, while other areas were added, such as an infirmary, a kitchen, and spaces for the staff in attendance. It was not until the 11th century that a certain spirit of research was re-established thanks to the influence of Arabic medicine. Again, it was the prominent figure of Avicenna who, along with the birth of hospitals, gave rise to the provision of healthcare in Baghdad. From there, the institution of centres of healthcare spread to the major cities of Mecca, Medina, and Cairo. In these cities, hospitals were generally built near mosques and under the patronage and authority of sultans. They were divided into specialist sections, including orphanages, libraries, and medical schools (Figures 9.1 and 9.2).

As already mentioned, Arabic medicine was incorporated into Christian medicine, particularly at the Salerno School, which substantially influenced the development of medical assistance. In Italy, the appearance of hospitals began to be documented in the 11th century.[22] They provided care to all categories of sick people, but some specialised in certain areas, particularly caring for patients with severe infections, while others assisted members of ecclesiastical organisations. Hygiene was poor, and patients with varying ailments were generally housed together. Dante included this scenario in the pestilential *bolgie* (rings) of the *Inferno*, where the damned are crowded together as a reference to hospitals in central Italy, particularly those affected by malaria:

> Qual dolor fora, se de li spedali
> di Valdichiana tra 'l luglio e 'l settembre
> e di Maremma e di Sardigna i mali
>
> fossero in una fossa tutti 'nsembre,
> tal era quivi, e tal puzzo n'usciva
> qual suol venir de le marcite membre.
> (*Inf.* XXIX, 46–51).

> [If the contagion of every hospital / in Valdichiana, from July until September, / and in the Maremma and Sardinia, were amassed / in one malarial ditch, such suffering / was in that place. And from it rose / the stench of festering limbs].

Figure 9.1 The interior of a medieval hospital. Illustration from ms Florence, Biblioteca Medicea Laurenziana, Gaddi 24, c. 247v. Reproduced by the kind permission of MiC – Ministero della Cultura. Any further reproduction by any means whatsoever is prohibited.

Figure 9.2 A medieval pharmacy and medical care. Illustrations from ms Bologna, Biblioteca Universitaria, 102, 2197, c. 492r. Reproduced by kind permission of the Biblioteca Universitaria. Any further reproduction by any means whatsoever is prohibited.

In addition to hospitals in the provinces of Val di Chiana, Maremma, and Sardinia, which were affected by malaria and undoubtedly unhealthy, other important Italian hospitals, such as the Florentine Santa Maria Nova, observed much healthier and better-organised rules. It was inaugurated on 23 June 1288 by Folco Portinari, the father of Beatrice, and had its own pharmacy and medicinal herb garden.[23] In addition, the care of patients followed the evangelic spirit and Franciscan rule. Santa Maria Nova was a model for many other European hospitals and continues to be an elite healthcare institution today; it still houses its ancient pharmacy, where some fine art treasures are held.[24]

Although there is no evidence that Dante attended hospitals in his lifetime, his active involvement in the public and military events of his native town may have put him in contact with institutions where sick or wounded people were hospitalised. This may also have contributed to his medical culture – or, at least, to his interest in diseases and human anatomy and physiology. The bloody Battle of Campaldino, which was fought by the Florentine Guelphs with their allies against the Ghibellines of Arezzo, where Dante participated as a *feditore* (i.e., a first-line assailant), resounds in *Purgatory* V in the tragic episode of Bonconte of Montefeltro.

When asked by Dante about the circumstances of his death, Bonconte replies:

> "Oh!", rispuos'elli, "a piè del Casentino
> traversa un'acqua c'ha nome l'Archiano,
> che sovra l'Ermo nasce in Apennino.
>
> Là 've 'l vocabol suo diventa vano,
> Arriva' io forato ne la gola,
> fuggendo a piede e sanguinando il piano.
>
> Quivi perdei la vista e la parola;
> nel nome di Maria fini', e quivi
> caddi, e rimase la mia carne sola".
> (*Purg.* V, 94–102)

> ["Ah", he replied, "at Casentino's border / runs a stream called Archiano / that springs above the Hermitage among the Apennines / To where its name is lost I made my way, / wounded in the throat, fleeing on foot, / and dripping blood across the plain. / There I lost sight and speech. / I ended on the name of Mary and there I fell, / and only my flesh remained"].

The first verse of the last tercet may originate from a personal experience of Dante, who, as a warrior, may have been aware that a man who suffers extensive bleeding may lose sight and speech immediately before his death. This supports our argument that even in an inquisitive and perceptive individual like Dante, tragic events such as battles could have been integrated into a form of medical culture

3 The Meanings of *nervi*

Dante incorporated medieval notions of neuroanatomy into his works and delved into both the central and peripheral nervous systems, as we have examined in previous studies, specifically in relation to Dante's observations about neurology and anatomy.[25] Given that the topic is examined in another chapter of this volume, we hereafter focus on related issues presented in the *Comedia*, namely symptoms linked to metal intoxication, cardiovascular stress, the physiology of phonation, human reproduction, and locomotion. First, however, we consider how Dante used a specific anatomical term – that is, *nerbo* or *nervo*, which designates the nerve – to describe various aspects of the nervous system.

As Dante reaches the Sixth circle of Hell, a component of the central nervous system is mentioned: the optic nerve, which is defined as "il nerbo del viso" (*Inf.* IX, 73–74: "the nerve of vision"). In the third ring of the seventh Circle, where sodomites are punished, Dante refers to "where he left his sin-stretched nerves" (*Inf.* XV, 114: "dove lasciò li mal protesi nervi"), using the word "nervi" to indicate the male organ. According to Gilson, this verse corresponds to the idea that the penis is closely related to the nervous system, as Albert the Great also argued.[26] Moreover, this verse may indicate a medieval awareness of the function of the vegetative nervous system.[27] In the *Comedia*, the term "nerbo" is also used to mean "tendon", that is, "the sinew of whose feet he held clutched" (*Inf.* XXI, 36: "e quei tenea de' piè ghermito 'l nerbo"). This demonstrates the confusion between these anatomical structures in the Middle Ages, which is a consequence of a similar uncertainty among authors of the classical Greek and Roman period.[28]

With reference to medical terminology, Dante's *Inferno* may have inspired neuroanatomical terminology used to designate one portion of the brain, the limbic system. As already mentioned, in Dante's masterpiece, Limbo indicates the first infernal circle, a dark and painful valley on the border of Hell inhabited by virtuous non-Christian spirits:

> de la valle d'abisso dolorosa
> che 'ntorno accoglie d'infiniti guai.
> Oscura e profonda era e nebulosa.
> (*Inf.* IV, 8–10)

> [the valley dolorous of the Abyss, / which gathers the deep sound of countless woes. / It was obscure, profound, and vaporous].

Similarly, the limbic system is an 'obscure' neurological structure consisting of a ring or a series of concentric rings that includes a set of neural circuits in the deepest and most ancient part of the forebrain (limbic lobe), which are linked to essential functions for the conservation of the species.[29] According to the French neurologist Jean-Didier Vincent, "like the Limbo of Christian [i.e., Dante's] mythology, the limbic system is the link between the neocortical sky and the reptilian hell".[30] Dante's exploration of anatomical themes within his works reflects the extent of medieval knowledge and the enduring influence of his poetic genius on subsequent fields, including neuroanatomy.

4 Neuropathological Conditions

The *Comedia* also contains intriguing references to symptoms and signs that could be associated with neurological diseases. One notable instance occurs in *Inferno* XXIV, which presents a vivid description of what appears to be an epileptic seizure. In the eighth circle of Hell, which is named Malebolge and designated for the punishment of thieves, Dante encounters a scene in which a snake bites one of the damned between the neck and shoulder, causing him to collapse. The subsequent depiction of the thief's recovery is likened to the awakening of an epileptic patient after a seizure:

> E qual è quel che cade, e non sa como,
> per forza di demon ch'a terra il tira,
> o d'altra oppilazion che lega l'omo,
>
> quando si leva, che 'ntorno si mira
> tutto smarrito de la grande angoscia
> ch'elli ha sofferta, e guardando sospira;
>
> tal era 'l peccator levato poscia.
> Oh potenza di Dio, quant'è severa,
> che cotai colpi per vendetta croscia!
> (*Inf.* XXIV, 112–118)

> [and just as one who faints, and knows not why / whether possessed by devils that pull him down /or seized by the sickness that causes men to fall / rises to his feet, and gazes round, / wholly bewildered by the breathless anguish / he has undergone, and as he looks, he sighs, / such did that sinner seem when he had risen. / O how stern it is, the power of God, / hurling such blows as it takes vengeance!].

The use of terms like *smarrito* (bewildered), *angoscia* (anguish) and *sospira* (he sighs) meticulously captures the post-ictal state of an epileptic individual, which has prompted some scholars to entertain the notion that Dante himself may have experienced seizures.[31] They include Cesare Lombroso, the famous Italian physician and criminologist, who suggested that Dante may have suffered from a form of epilepsy characterised by frequent spells and visions.[32] While these are mere speculations, the verses in question shed light on medieval beliefs about the aetiology of epilepsy. In ancient times, epilepsy was often attributed to a demonic entity that possessed an individual. However, in explaining the pathogenesis of this disorder, Dante employed a technical term from medieval medicine: "oppilazion", which signifies the obstruction of the brain ventricles, prevents the proper movement of bodily humours, and results in fainting and trembling.[33]

Moreover, the *Comedia* may provide an early depiction of the effects of metal intoxication on both the central and peripheral nervous systems. In the medieval era, the use of metals was widespread, and there was an awareness that certain metals could have detrimental effects on human health. Amid counterfeiters in Hell, Dante encounters alchemists who suffer from limb paralysis: "who had not strength enough to lift their bodies" (*Inf.* XIX, 72: "che non potean levar le lor persone"). Conversely, some counterfeiters exhibit signs of irritability and aggression. For instance, Gianni Schicchi is described (*Inf.* XXX, 33) as *rabbioso* (rabid). Some scholars have proposed that both the physical conditions (such as palsy) and behavioural changes (mood swings) observed in alchemists could be attributed to the neurotoxic effects of metals.[34] This portrayal in Dante's verses is compatible with historical knowledge of the potential hazards associated with metal exposure, which highlight a nuanced understanding of the impact of such substances on the human nervous system.

Dante's experiences of loss of consciousness, hallucinations, and fainting are pervasive throughout the *Comedia*. Some modern scholars have proposed that, while he attributes these recurring episodes to love, passion, fear, anguish, and other emotions, they may actually signify a genuine pathological condition that Dante may have endured since childhood. Specifically, according to Plazzi, narcolepsy emerges as a clinical entity that aptly encompasses the various symptoms that afflict Dante during his journey through the afterlife.[35] These include cataplexy, hallucinations, sleep paralysis (*Inf.* IV, 1–4), episodes of muscle weakness (*Inf.* I, 52), and falls, which are all consistently triggered by intense emotions (*Inf.* V, 142). In the initial verses of his masterpiece, Dante presents himself as "full of sleep" (*Inf.* I, 11: "pien di sonno") and disoriented; this suggests that his entire imaginative journey could be figuratively interpreted from a medical perspective, as the result of a hypnagogic or hypnopompic hallucination in a subject afflicted by hypersomnia.

The *Inferno* also contains descriptions of psychiatric disorders, notably depression, which is Dante's melancholia. This is not considered a sin, but a diseased state (i.e. a fitting punishment for souls who transgress God's laws and succumb to sin in their earthly lives). According to Widmer, "Dante Alighieri's use of terms and descriptions so strongly associated with melancholia suggests that it is this disease and its emotional state that epitomized the environment of the Inferno".[36] Weakness, limb paralysis, rigidity, and heaviness, experienced by many denizens of Hell, could be interpreted as expressions of psychomotor dysfunction that result from an excess of black bile in the bodies of melancholic souls. Dante's portrayal of characters who are afflicted by depression suggests a secret sympathy for these tormented souls, that may be rooted in his personal experiences of sadness and grief.

Dante penned the *Comedia* during his exile from Florence, and his emotional entanglement with suffering souls has been analysed as a form of countertransference by scholars who have explored Dante's work from a psychoanalytic viewpoint. Some also view Dante's approach to psychological disorders as a medieval precursor to psychoanalytic techniques, integrating and revising ancient models of psyche and personhood.[37]

5 Dante's Cardiovascular Perspectives

Dante's *Comedia* includes a rich list of anatomical and physiological references to the circulatory system, which reveals some understanding of related diseases and their remedies.[38] These observations likely drew inspiration from the works of Aristotle and Galen, for whom the heart held a pivotal role as the seat of all passions and the central organ that orchestrates the functions of the entire body "where nature is intent on all the members" (*Purg.* XXV, 60: "dove natura a tutte membra intende"). Moreover, the prevailing belief was that venous blood not only nourished but also contributed to the formation of the body's soft tissues and viscera.[39] The vital role of blood is exemplified, for instance, in the episode of Bonconte from Montefeltro (*Purg.* V, 74), which correlates the loss of blood with the cessation of life. Moreover, in the poem *Il Fiore* (*The Flower*), which is often attributed to Dante, haemorrhage is portrayed as inducing weariness: "When I felt myself so drained / because of all the blood I'd lost" (*Fiore* II, 2: "Sententomi ismagato / del molto sangue ch'io avea perduto").

An interesting observation by Dante showcases a sort of pioneering understanding of the circulatory system in its description of the pallor of a lover that results from blood flowing from the veins to the heart:

> E 'l sangue ch'è per le vene disperso
> fuggendo corre verso
> lo cor, che 'l chiama, ond'io rimango bianco.
> (*Rime* CIII, VI, 45–47)
>
> [and the blood / that was dispersed through my veins / runs fleeing back to the heart / that summons it, so that I am left white].[40]

However, this depiction is better interpreted as a poetic metaphor for love, which was believed to draw all of the body's blood towards the supposed source, the heart.

Notably, in the *Comedia*, Dante does not appear to distinguish between veins and arteries and describes the blood flowing from the veins as either brown (*Inf.* XIII, 34: "sangue bruno") or flaming red (*Purg.* IX, 101–102: "fiammeggiante"), which underscores the unclear differentiation between arterial and venous blood in medieval physiology.[41] His depiction of the heart as the origin of sperm further highlights his limited anatomical and physiological understanding of the cardiovascular system. However, his descriptions of cardiovascular symptoms exhibit a greater degree of accuracy. For example, an instance of orthopnoea is evident in Dante's depiction of the ailing woman in Purgatory, to whom he compares Florence:

vedrai te somigliante a quella inferma
che non può trovar posa in su le piume,
ma con dar volta suo dolore scherma
(*Purg*. VI, 149–151)

[Thou shalt behold thyself like a sick woman, / Who cannot find repose upon her down, / But by her tossing wardeth off her pain].

Furthermore, shortness of breath and dyspnoea on exertion are described when Dante and his guide, the Roman poet Virgil, traverse the circles of Hell (*Inf*. XXIV, 43–45; *Inf*. XXXIV, 82–83) or ascend the mountain of Purgatory (*Purg*. IV, 115–116; *Purg*. XXIV, 70–72). Although these symptoms may not be exclusively linked to cardiovascular diseases, their portrayal suggests a capacity for associating dyspnoea with physical fatigue, potentially of cardiovascular origin. Some have attempted to interpret Dante's descriptions as signifying specific cardiovascular pathologies, including the words spoken by Count Ugolino della Gherardesca in response to Dante's prompting him to recall the dreadful deaths of his sons, whose corpses he is supposed to have devoured due to hunger. Ugolino's words have been interpreted as a pioneering reference to angina pectoris: "Thou wilt that I renew / the desperate grief, which wrings my heart already" (*Inf*. XXXIII, 4–6: "tu vuo ch'io rinovelli / disperato, dolo che cor mi preme").[42]

Another intriguing description of pathologies that may be associated with cardiovascular diseases relates to Mastro Adamo, who was condemned as a money counterfeiter, and generally believed by Dante scholars to suffer from ascites[43]:

Io vidi un, fatto a guisa di leuto
pur ch'elli avesse avuta l'anguinaia
tronca da l'altro che l'uomo ha forcuto
(*Inf*. XXX, 49–51)

[I saw one man made in fashion of a lute, / If he had only had the groin cut off / Just at the point at which a man is forked].

Though Mastro Adamo's ascites are commonly attributed to alcoholic liver disease, cardiovascular causes such as congestive heart failure, constrictive pericarditis, and valvular heart disease should also be considered as causes of the "thirsty dropsy" (v. 52: "la grave idropesì") that affects him.[44] It is known in modern medicine that individuals with chronic heart failure often experience severe thirst. The metaphorical association between thirst and dropsy has deep roots in the works of Christian authors, such as St. Augustine and Caesarius of Arles. Augustine, for instance, metaphorically linked the pursuit of wealth to the symptoms of dropsy: "When somebody wants to become rich, he desires, and burns, and becomes thirsty; and just like a man affected by the disease of dropsy, the more he drinks, the more he gets thirsty".[45]

This striking similarity between a physical ailment and the spiritual condition of greed, suggests that a covetous individual is akin to one with dropsy in the heart. Both conditions, whether of the soul or body, exacerbate the problem when the desired goal (money or water) is pursued. Dante vividly describes Mastro Adamo's insatiable thirst and likened it to water bloating him out (v. 126: "ché s'i' ho sete e omor mi rinfarcia"). The damned soul experiences eternal thirst as a consequence of his sins, enduring suffering as part of his "contrapasso", a "punishment by suffering the opposite"[46]). Therefore, this portrayal should not merely be interpreted as an actual disease suffered by the character, and suggestions of specific pathological mechanisms (e.g., "hepatic or renal toxic effects of copper or other non-noble metals") should be considered highly speculative.[47]

Lastly, Dante appears to have a propensity for fainting throughout his extended journey in Hell and Purgatory. In Hell, he faints twice: first, during the passage across the Acheron "and as a man whom sleep hath seized I fell" (*Inf.* III, 136: "e caddi come l'uom cui sonno piglia") and second in the circle of the lustful "I swooned away as if I had been dying. / And fell, even as a dead body fall" (*Inf.* V, 141–142: "io venni men così com'io morisse. / E caddi come corpo morto cade"). In addition, Dante collapses one last time after being overcome with emotion at the edge of the Purgatory mountain: "Such knowledge of my fault was gnawing at my heart / that I was overcome" (*Purg.* XXXI, 89: "Tanta riconoscenza il cor mi morse, / chio caddi vinto"). Finally, the *Vita Nuova* describes Dante's predisposition to emotional syncopes, as exemplified by the following report:

> Però quando mi tolle sì 'l valore,
> che li spiriti par che fuggan via,
> allor sente la frale anima mia
> tanta dolcezza, che 'l viso ne smore
> (Vita Nuova, XXVII)
>
> [And so when he takes away my courage, and my spirits seem to fly away, then I feel throughout my soul such sweetness that my face pales].[48]

Some scholars have suggested that this recurrent fainting could be the manifestation of a neurological disorder, possibly epilepsy or narcolepsy.[49] Dante wrote that he made his metaphorical journey to the afterlife when he was only 35 years old, which suggests that he indeed suffered from a neurological disorder, as he appears to have been healthy during his youth in every current description. Scholars have postulated that the loss of consciousness reported in the fifth Canto of *Inferno* could be the manifestation of an emotional syncope.[50] Additional information to support this hypothesis may be gleaned by further analysing Dante's description of his imaginative journey into the afterlife. According to the scientific literature, two other factors mainly contribute to episodes of emotional syncope: reduced blood volume and vasodilation, which worsen the effects of hypovolemia on cerebral perfusion.

Although we cannot speculate about Dante's diet and salt intake, some hypotheses can be made about other factors that influenced his blood volume and ability to maintain wakefulness.[51] According to the poet, his journey in Hell began on the night of 8 April 1300 and ended in Paradise one week later. Apparently, the pilgrim Dante never has an opportunity to stop and drink water or eat during his entire passage through Hell, which lasts around 48 hours, and does not drink any water in Purgatory. Thus, it can be concluded that he did not drink or eat while walking and experienced heavy emotional distress for five consecutive days. After Dante's last collapse in Purgatory, his guide Beatrice throws him into the Lethe River, where he drinks for the first time on his long journey:

> La bella donna ne le braccia aprissi;
> abbracciommi la testa e mi sommerse
> ove convenne ch'io l'acqua inghiottissi.
> (*Purg.* XXXI, 100–102)
>
> [The beautiful lady opened wide her arms, / embraced my head, and plunged me underneath, / where I was forced to swallow of the water].

A few hours later, Dante drinks the sweet water of the Eunoe River: "the drink that never would have sated me" (*Purg.* XXXIII, 138: "lo dolce ber che mai non m'avria sazio"). Subsequently, his fainting ceases. Another factor that may contribute to the distress and multiple syncopes suffered by the poet is the temperature that he endured as a pilgrim to Hell.

While the details surrounding Dante's experiences of abstaining from drinking, grappling with emotional distress, and enduring elevated temperatures during his journey through the afterlife are drawn from the poetic narrative, it is noteworthy that various historical sources attested to the medieval physicians' recognition of the adverse effects of such conditions on the human body, including a propensity for fainting. Therefore, these deductions, which are rooted in empirical observations and common sense, are also underpinned by the prevailing medical principles of the time.[52]

6 Mechanisms and Functions: Human Reproduction, Phonation and Locomotion

The preceding paragraphs delve into how Dante's *Comedia* can offer valuable insights into the anatomical and physiological understanding of the cardiovascular and nervous systems during the Middle Ages. However, Dante also showed a remarkable insight into the physiological mechanisms that underpin other crucial aspects of human life, such as reproduction, phonation, and locomotion, a knowledge inherited from medieval medicine.

The medieval understanding of human reproduction and embryogenesis is vividly depicted in certain passages of the *Comedia*.[53] Departing from the fifth *Cornice* of Purgatory, Dante encounters the soul of the Latin poet Statius, who expounds on how the human body is generated in the uterus. In reference to perfect blood coming from the heart, Statius says:

> Ancor digesto, scende ov'è più bello
> tacer che dire; e quindi poscia geme
> sovr'altrui sangue in natural vasello.
>
> Ivi s'accoglie l'uno e l'altro insieme,
> l'un disposto a patire, e l'altro a fare
> per lo perfetto loco onde si preme;
>
> e, giunto lui, comincia ad operare
> coagulando prima, e poi avviva
> ciò che per sua matera fé constare.
> (*Purg*. XXV, 43–51)

> [Yet more concocted [the blood] descends, / where shame forbids to mention [testicles]: and from thence distils / in a natural vessel [uterus] on another's blood [menstruation]. / There each unites together; / one disposed to endure, to act the other, / through that power derived from whence it came; / and being met, [sperm] begins to work, / coagulating first; then vivifies / what its own substance makes].

These verses encapsulate medieval ideas about the genesis of life that were influenced by Hippocratic and Aristotelian theories, which gained renewed elaboration in medieval philosophy during the 13th century.[54] According to these theories, reproductive function is intricately linked to nutrition, where sperm is perceived as a residue of blood derived through the process of nourishment. Following ingestion, food undergoes a transformation into chyle in the intestinal tract; later, in the liver and heart, it becomes blood. In men, this blood, as it filters through various organs, undergoes purification and reaches the testicles, where it is ultimately transformed into sperm. Aquinas asserted that "the blood [i.e., sperm], that is prepared for conception through digestion, is purer and more perfect than other blood".[55] The union of sperm with menstrual blood in the uterus, which is subject to the influence of male semen, results in the formation of a clot that then develops into a living being. Thus, fertilisation was perceived as the creation of a clot that combines the active male principle (sperm) with the passive female principle (menstruation).[56]

In the aforementioned verses, Dante refers to the uterus as a "natural vasello" (natural vessel), which emphasises the woman's passive role in the process. This dichotomy between an active male role and a passive female role in human conception was derived from Aristotelian ideas and articulated by Dante in *Convivio* IV, xxxi, 4–5. In the latter, the poet posits that male semen imparts intellect, life, and spirit to the newly forming organism, while the female part plays a purely passive role; in other words, the uterus is merely designed to receive and nurture the life brought forth by sperm.[57]

A great attention to detail also emerges in Dante's portrayal of the mechanisms underlying human phonation.[58] According to his anatomical perspective, the primary organ responsible for speech is the tongue. This belief is evident in various descriptions throughout his works, which reflect the cultural and educational influences on his understanding of anatomy. For instance, in *Inferno* XXV, 133, he observes the transformation of thieves' souls, describing one thief's tongue as "fit for speech" ("presta / prima a parlar"). A similar analogy is drawn in *Inferno* XXVI, 89, where Dante encounters two fraudulent counsellors, Ulysses and Diomedes, in the eighth *bolgia*; they are imprisoned in a flame that he compares to a tongue: "Moving as if it were the tongue that spake it uttered forth a voice and said" ("indi la cima qua e là menando, / come fosse la lingua che parlasse, / gittò voce di fuori e disse"). However, Dante also highlighted the role of the throat in phonation. In *Inferno* XXIII, when he encounters the hypocrites adorned in golden clothes made of lead, two damned souls recognise him by the movement of his throat. Dante's breathing, observed through the motion of his larynx rather than his nostrils, signifies his living state and exemption from the 'heavy stole' of death. This suggests that it is the tongue that governs phonation, while the throat is mostly associated with respiration.[59] Nevertheless, a passage in *Inferno*, XXVIII, challenges this assumption. In the ninth *bolgia*, where sowers of discord and scandal within the papacy are punished, Dante meets Curio, a Roman soldier who influenced Caesar in his crossing of the Rubicon, leading to civil war. Curio has his throat cut, which renders him unable to speak:

> O how bewildered unto me appeared,
> With tongue asunder in his windpipe slit,
> Curio, who in speaking was so bold!
> (*Inf.* XXVIII, 100–102)

> [Oh quanto mi pareva sbigottito / con la lingua tagliata ne la strozza / Curïo, ch'a dir fu così ardito!].

In the above passage, the larynx is acknowledged as crucial not only for breathing but also for speech. Dante had previously described the throat of another one of the damned, Pier da Medicina, as a pipe (*canna*), which emphasises his understanding of laryngeal anatomy. Pier, with only a hole in his throat, can speak, whereas Curio cannot. While Dante may not have been aware of vocal cords, his work implies a recognition of the role of the throat in phonation through the metaphorical association of the tongue with speech in some instances, such as when Virgil advises Dante to "take heed that thy tongue restrain itself" (*Inf.* XXVI, 72: "ma fa che la tua lingua si sostegna").

Finally, Dante's references to musculoskeletal pathologies and conditions related to the locomotor system are illustrated by his employment of anatomical terms to denote musculoskeletal components, such as "omero" (humerus) to refer to the arm (*Purg.* IV, 120; *Purg.* XXVI, 4; *Purg.* XVI, 9), "calcagne" (calcaneus) to refer to the feet (*Purg.* XIX, 61), and "giunture" to generically refer to the joints (*Purg.* XXVI, 57).[60] Other terms suggest a rudimentary knowledge of spinal pathophysiology. For instance, Dante likely understood that cervical trauma could result in permanent or temporary paralysis. Two references to lower back pain associated with carrying heavy weights underline a correlation familiar to modern physicians in the context of chronic lower back pain and spinal pathology. In Purgatory, penitents from pride bear heavy stones on their backs with pain and affliction and display varying degrees of contraction and endurance:

> Vero è che più e meno eran contratti
> secondo ch'avien più e meno a dosso;
> e qual più pazïenza avea ne li atti,
> piangendo parea dicer: 'Più non posso'.
>
> (*Purg.* X, 136–139)
>
> [Each, as his back was laden, came indeed / or more or less contracted; and it appear'd / as he, who show'd most patience in his look, / wailing exclaimed: 'I can endure no more'].

Similarly, in Hell, fortune tellers are burdened with exceedingly heavy leaden hoods on their heads, which induces weeping and fatigue (*Inf.* XXIII, 60: "gravi tanto") and provides a realistic portrayal of this affliction.

Notes

1 Corsini, "Del Dottor Dante Alighieri"; Ballantyne, "Was Dante a Doctor?".
2 Lippi, *La Divina Comedia*; Boccaccio, *Trattatello*.
3 Tobino, *Biondo Era*, 12.
4 For a recent outlook on Dante's presence in Bologna, with detailed information on the school of medicine and related bibliography see Tabarroni, "Ambienti culturali prossimi a Dante".
5 Asín Palacios, *Dante e l'Islam*.
6 See Casagrande and Fioravanti, *La filosofia in Italia al tempo di Dante*, 13 and 22, note 10. Manfred's gift was accompanied by a beautiful letter inspired to the precepts of the *ars dictaminis* (the art of writing). The letter was originally attributed to Pier delle Vigne, the secretary and notary of Frederick II. Replying to Dante, who wishes to know his identity, the suicide says of himself: "I am who

held both keys to Frederick's heart" (*Io son colui che tenni ambo le chiavi / del cor di Federigo*: *Inf.* XV, 58–59). Manfred, in turn, is the protagonist of one of the most famous episodes of Dante's *Purgatorio* (III, 103–145).
7 Bernabeo, "La scienza medica in Dante", 30.
8 On Alderotti and the medicine in Bologna, see Siraisi, *Taddeo Alderotti*; Casagrande and Fioravanti, *La filosofia in Italia al tempo di Dante.*
9 Alderotti, *Expositiones.*
10 Di Matteo, Tarabella, Filardo, Mosca, Lo Presti, Viganò, Tomba, and Marcacci, "Art in Science".
11 Fughelli and Maraldi, "Mondino de' Liuzzi"; Sandor, Dumitrascu, Bojita, and Dumitriascu, "Medicine and Pharmacy". In *Inf.* XXXII, 127–129, Dante uses medical terms close to those present in Mondino's anatomical book.
12 Lippi, *Tra Ipocràte e Galieno.*
13 Examples of this portrayal are in the Strozziniano codex 174 of the Laurentian library; the chapel of Giotto's Palazzo del Bargello; the Uffizi Gallery by Andrea del Castagno, the Duomo of Florence by Domenico di Michelino, the Minneapolis Institute of Art with Portrait of six Tuscan poets by Giorgio Vasari.
14 Barbi, "Dante e l'arte dei medici", 382.
15 Barbera, *Atlante Dantesco*, 85.
16 See the contribution by Francesca Galli in this volume.
17 Premuda, "Aspetti medici".
18 See Tabarroni, "Ambienti culturali prossimi a Dante".
19 De Renzi, "La Medicina in Italia, 533–544.
20 Riesman, "A Physician in the Papal Chair"; Martins, "A Brief Biography".
21 See Pasi, "*Dante, i medici*, 11–14"; Lunedei, "Dante, la medicina".
22 Castiglioni, "La medicina ai tempi e nell'opera di Dante"; Li Calzi, Fontana, and Sandolo, *Per una storia.*
23 Ruffino, "L'antica spezieria".
24 Diana, "La spezieria dell'ospedale".
25 Riva, Bellani, Tremolizzo, Lorusso, Ferrarese, and Cesana, "The Neurologist".
26 Gilson, "The Anatomy".
27 Gilson, "The Anatomy".
28 Riva, Bellani, Tremolizzo, Lorusso, Ferrarese, and Cesana, "The Neurologist".
29 Olry and Haines, "NEUROwords".
30 Vincent, *Biologie des passions*, 134.
31 Riva, Bellani, Tremolizzo, Lorusso, Ferrarese, and Cesana, "The Neurologist".
32 Lombroso, "Le nevrosi"; Plazzi, "Dante's Description".
33 Lustig, *Dizionario*, 316.
34 Lucchini, Riva, Sironi, and Porro, "Torvis oculis".
35 Plazzi, "Dante's Description".
36 Widmer, "Black Bile".
37 Chessick, "Dante's Divine Comedy Revisited"; Gomez, "Some Psychoanalytic Thoughts".
38 Riva, Cambioli, Castagna, Cianci, Varrenti, Giannattasio, and Cesana, "Dante and Cardiology".
39 Gilson, "The Anatomy".
40 Riva, Cambioli, Castagna, Cianci, Varrenti, Giannattasio, and Cesana, "Dante and Cardiology".
41 Premuda, "Introduction".
42 Riva, Cambioli, Castagna, Cianci, Varrenti, Giannattasio, and Cesana, "Dante and Cardiology".
43 Bianucci, Perciaccante, Charlier, Appenzeller, and Lippi, "Mastro Adamo".
44 Riva, Cesana, Achilli, and Scordo, "The 'Thirsty Dropsy'".

45 Augustine, *Sermones*: "Volendo dives esse, desiderat, aestuat, sitit; et tamquam hydropisis morbo, plus bibendo, plus sitit".
46 Smith, "So the Satiated Man Hungers".
47 Bianucci, Perciaccante, Charlier, Appenzeller, and Lippi, "Mastro Adamo".
48 Riva, Cambioli, Castagna, Cianci, Varrenti, Giannattasio, and Cesana, "Dante and Cardiology".
49 Lippi, *La Divina Commedia*; Ballantyne, "Was Dante a Doctor?".
50 Estañol, Delgado, Jiménez-Mayo, and Sentíes-Madrid, "An Early Literary Description".
51 Riva, Cambioli, Castagna, Cianci, Varrenti, Giannattasio, and Cesana, "Dante and Cardiology".
52 See again Riva, Cambioli, Castagna, Cianci, Varrenti, Giannattasio, and Cesana, "Dante and Cardiology".
53 This "scientific digression" on embryology has a wide bibliography. See for instance Ureni, "Human Generation". Here we follow mainly Locatelli, Nespoli and Riva, "Dante and Reproduction".
54 Locatelli, Nespoli, and Riva, "Dante and Reproduction".
55 In his *Summa Theologica* (III, 31): "sanguis, qui digestione quadam est preparatus ad conceptum, est purior et perfectior alio sanguine".
56 Locatelli, Nespoli, and Riva, "Dante and Reproduction".
57 Locatelli, Nespoli, and Riva, "Dante and Reproduction".
58 Borraccini, Marinini, and Riva, "Medieval Interpretation of Phonation".
59 Borraccini, Marinini, and Riva, "Medieval Interpretation of Phonation".
60 Manzini, Maurina, and Riva, "Dante and Musculoskeletal Diseases".

Bibliography

Primary Sources

Augustine. *Sermones de vetere Testamento (1–50)*. Edited by Cirille Lambot. Turnhout: Brepols, 1997.

Giovanni Boccaccio. *Trattatello in laude di Dante*. Milan: Garzanti, 2007.

Taddeo Alderotti. *Expositiones Aphorismorum Hippocratis*. Venice: Giunta, 1527.

Secondary Sources

Asín Palacios, Miguel. *Dante e l'Islam. L'escatologia islamica nella Divina Commedia*. Turin: Pratica, 1993.

Ballantyne, John William. "Was Dante a Doctor?". *British Medical Journal* 1 (1910): 331–333; 395–396.

Barbera, Gianluca. *Atlante Dantesco. I luoghi di Dante e della Divina Commedia*. Milan: Rizzoli, 2022.

Barbi, Michele. "Dante e l'arte dei Medici e Speziali". In. *Problemi di critica dantesca. Seconda serie 1920–1937*, edited by Michele Barbi, 379–384. Florence: Sansoni, 1975.

Bernabeo, Raffaele A. "La scienza medica in Dante". In *Atti del Convegno su Dante e l'enciclopedia delle scienze*, edited by Emilio Pasquini, 29–39. Bologna: Clueb, 1991.

Bianucci, R., A. Perciaccante, P. Charlier, O. Appenzeller, and Donatella Lippi. "Mastro Adamo, the Counterfeiter of Coins, Had Cirrhosis as Described in Dante's

Inferno (13th Century Florence). *European Journal of Internal Medicine* 39 (2017), e35–e36.

Borraccini, M., M. Marinini, and Michele A. Riva. “Medieval Interpretation of Phonation Through the Divine Comedy by Dante Alighieri”. *Otolaryngol Head Neck Surgery* 167/1 (2022): 109–111.

Casagrande, Carla and Gianfranco Fioravanti. Eds. *La filosofia in Italia al tempo di Dante*. Bologna: Il Mulino, 2016.

Castiglioni, Arturo. “La medicina ai tempi e nell’opera di Dante”. *Archivio di Storia della scienza* 3 (1922), 211–236.

Chessick, Richard D. “Dante’s Divine Comedy Revisited: What Can Modern Psychoanalysts Learn from a Medieval ‘Psychoanalysis’?”. *Journal of The American Academy of Psychoanalysis* 29/2 (2001): 281–304.

Corsini, Guido. “Del Dottor Dante Alighieri”. *Giornale del Centenario di Dante Allighieri* 2 (1864): 11–15.

De Renzi, Salvatore. “La Medicina in Italia ai tempi di Dante”. In *Dante e il suo secolo*, 533–544. Florence: Mariano Cellini, 1865.

Di Matteo, Bernardo, Vittorio Tarabella, Giuseppe Filardo, Massimiliano Mosca, Mirco Lo Presti, Anna Viganò, Patrizia Tomba, and Maurilio Marcacci. “Art in Science. Mondino de’ Liuzzi. The Restorer of Anatomy”. *Clinical Orhtopaedics and Related Researches* 475 (2017): 1791–1795.

Diana, Esther. “La spezieria dell’ospedale di Santa Maria Nuova tra scienza ed arte”. *Atti e memorie. Rivista di Storia della Farmacia* (2016): 223–230.

Estañol, Bruno, Guillermo Delgado, Eduardo Jiménez-Mayo, and Horacio Sentíes-Madrid. “An Early Literary Description of Emotional Syncope in the Fifth Canto of Dante Alighieri’s Commedia: ‘E caddi come corpo morto cade’”. *Journal of Cardiovascular Medicine* 15 (2014): 604–608.

Fughelli, Patrizia and Maraldi, Elisa. “Mondino de’ Liuzzi e Dante Alighieri. Frequentazioni mediche bolognesi?”. *Griseldaonline. Centro Studi Camporesi* (2012): 1–7. https://centri.unibo.

Gilson, Simon A. “The Anatomy and Physiology of the Human Body in the *Comedia*”. In *Dante and the Human Body: Eight Essays*, edited by John C. Barnes and Jennifer Petrie, 11–42. Dublin: UCD Foundation for Italian Studies/Four Courts Press, 2007.

Gomez, Efrain A. “Some Psychoanalytic Thoughts about King Lear, Dante, and Don Quixote”. *Journal of The American Academy of Psychoanalysis* 14/4 (1986): 545–556.

Li Calzi, Epifanio, Stefano Fontana, and Alessandra Sandolo. *Per una storia dell’architettura ospedaliera*. Milan: Maggioli, 2002.

Lippi, Donatella. *Dante tra Ipocràte e Galieno. Il lessico della medicina nella Commedia*. Florence: Angelo Pontercorboli Editore, 2021.

Lippi, Donatella. *La Divina Commedia. Con note storico-mediche*. Fidenza: Mattioli, 1885, 2009.

Locatelli, Luca, Antonella Nespoli, and Michela A. Riva. “Dante and Human Reproduction: Concept of Embryogenesis in the Middle Ages”. *Romanian Journal of Morphology and Embryology* 56/1 (2015): 325–326.

Lombroso, Cesare. “Le nevrosi in Dante e Michelangelo”. *Archivio di Psichiatria* 15 (1894): 126–132.

Lucchini, Roberto G., Michele A. Riva, Vittorio A. Sironi, and Alessandro Porro. “Torvis oculis: Occupational Roots of Behavioural Neurotoxicology in the Last Two Centuries and Beyond”. *Neurotoxicology* 33 (2012): 652–659.

Lunedei, Antonio. "Dante, la medicina e i medici del suo tempo". *Rivista Critica di Clinica Medica* 5 (1967): 5–29.
Lustig, Renzo. *Dizionario di terminologia medica*. Milan: Società editrice libraria, 1927.
Manzini, Francesco, Matteo Maurina, and Michele A. Riva. "Dante and Musculoskeletal Diseases: Anatomy and Pathophysiology of Orthopaedic and Spinal Disorders in the Middle Ages". *Spine (Phila Pa 1976)* 40/18 (2015):1467–1469.
Martins, Paulo Nuno. "A Brief Biography of the Pope Pedro Hispano: A Portuguese Philosopher and Physician of the Christian World". *International Journal of Philosophy and Theology* 5 (2017): 44–46.
Olry, Régis and Duane E. Haines. "NEUROwords: From Dante Alighieri's First Circle to Paul Donald MacLean's Limbic System". *Journal of the History of the Neurosciences* 14 (2005): 368–370.
Pasi, Romano. *Dante, i medici e la medicina*. Ravenna: Essegi, 1996.
Plazzi, Giuseppe. "Dante's Description of Narcolepsy". *Sleep Medicine* 14 (2013): 1221–1223.
Premuda, Loris. "Introduction" to Guglielmo Harvey. *De Motu Cordis*, edited by Loris Premuda, 61–74. Milan: Industrie Grafiche Nicola Moneta, 1957.
Premuda, Loris. "Aspetti medici nell'opera di Dante". *Lo Smeraldo* 19/3 (1965): 31–36.
Riesman, David. "A Physician in the Papal Chair". *Annals of Medical History* 5 (1923): 291–300.
Riva, Michele A., Iacopo Bellani, Lucio Tremolizzo, Lorenzo Lorusso, Carlo Ferrarese, and Giancarlo Cesana. "The Neurologist in Dante's Inferno". *European Journal of Neurology* 73/5–6 (2015): 278–282.
Riva, Michele A., L. Cambioli, F. Castagna, N. Cianci, M. Varrenti, C. Giannattasio, and Giancarlo Cesana. "Dante and Cardiology: Physiopathology and Clinical Features of Cardiovascular Diseases in the Middle Ages". *International Journal of Cardiology* 181 (2015): 317–319.
Riva, Michele A., Giancarlo Cesana, F. Achilli, and F. Scordo. "The 'Thirsty Dropsy': Early Descriptions in Medical and Non-Medical Authors of Thirst as Symptom of Chronic Heart Failure". *International Journal of Cardiology* 245 (2017): 187–189.
Ruffino, I. "L'antica spezieria". In *Santa Maria Nuova attraverso i secoli*, edited by Giancarlo Landini, 239–246. Florence: Edizioni Polistampa, 2017.
Sandor, Vlaicu, Dinu I. Dumitrascu, Marius T. Bojita, and Dan L. Dumitriasco. "Medicine and Pharmacy in the Work of Dante Alighieri (1265–1321)". *Medicine and Pharmacy Reports* 95 (2022): 218–224.
Siraisi, Nancy G. *Taddeo Alderotti and His Pupils: Two Generations of Italian Medical Learning*. Princeton, NJ: Princeton University Press, 2019.
Smith, James L. "'So the Satiated Man Hungers, the Drunken Thirst': The Medieval Rhetorical Topos of Spiritual Nutrition". *Limina* 20 (2015): 1–17.
Tabarroni, Andrea. "Ambienti culturali prossimi a Dante nell'esilio: lo Studio bolognese di arti e medicina". In *Dante fra il settecentocinquantenario della nascita (2015) e il settecentenario della morte (2021). Atti delle Celebrazioni in Senato, del Forum e del Convegno internazionale di Roma: maggio-ottobre 2015*, edited by Enrico Malato and Andrea Mazzucchi, vol. 1, 327–348. Rome: Salerno editrice, 2016.
Tobino, Mario. *Biondo era e bello*. Milan: Mondadori, 1974.

Ureni, Paola. "Human Generation, Memory and Poetic Creation: From the *Purgatorio* to the *Paradiso*". *Quaderni d'italianistica* 31/2 (2010), 9–34.

Vincent, Jean-Didier. *Biologie des passions*. Paris: Odile Jacob, 1986.

Widmer, David A. J. "Black Bile and Psychomotor Retardation: Shades of Melancholia in Dante's Inferno". *Journal of the History of the Neurosciences* 13 (2004): 91–101.

10 Dante, Neurologist and Neuroanatomist

Evidence from the *Divine Comedy*

Donatella Lippi, Raffaella Bianucci, Elena Varotto, Francesco Arba and Francesco Maria Galassi

1 A Brief State of the Art on Ancient and Medieval Neurological Knowledge

Many biographical elements confirm the link between Dante and the world of medicine, which is also widely reflected in his literary work.[1] In this perspective, Dante's acquaintance with the neurological knowledge of his day appears to be particularly relevant. To frame this knowledge in its historical context, it is useful to briefly recall the state of the art of the ancient medical conceptions about the notions that we call neurologic and diseases that affect the nervous system. Hippocrates of Coos (5th century BCE) elaborated a physio-pathological system based on the four humours: blood, phlegm, yellow bile, and black bile, originating, respectively, in the heart, brain, liver, and spleen.[2] Within the framework of this philosophical and medical conception, health is determined by a balanced mix of humours. Indeed, according to Hippocrates, a disease such as epilepsy is the result of an excess of phlegm in the brain. Moreover, in Hippocratic and subsequent Classical medicine, epilepsy was also linked to melancholia ("chronic depression") caused by black bile produced by the spleen.[3]

Although Hippocrates and other physicians, such as Alcmaeon of Croton (fl. 5th century BCE), had previously identified the brain as the seat of thought, it was Aristotle who ascribed to this organ other roles, such as that of moderating blood temperature. Indeed, for Aristotle, while humans alone have mind and reason, several animals have a brain. This means that this organ has to serve to other functions than thinking. However, Aristotle systematically denied that the brain controls sensation and movement and is the centre of emotional functions. All these neurological functions were instead located in a totally different organ, the heart.[4] Many centuries later, one of the most influential Roman physicians, Galen of Pergamum (*c.*129–216 CE), physician of Emperor Marcus Aurelius, offered a much more elaborated description of the brain, likely based on human autopsies. He enumerated seven pairs of cranial nerves (currently known to be 12 pairs).[5] Moreover, his studies on the animals led him to confirm that damage in the section of the spinal cord, located between the first and second vertebrae, resulted in the

DOI: 10.4324/9781003412601-11

animal's immediate death, whereas an injury in the lumbosacral spine was accompanied by paralysis of the lower limbs and sphincters.[6] He was also the first to introduce the term "migraine" (from the Greek ἡμικρανία, litterally "half head", from which "pain on one side of the head") to underline the one-sidedness characteristic of this form of headache.[7] Aretaeus of Cappadocia (fl. 2nd century CE) offered careful clinical descriptions of various neurological ailments including epilepsy, paralysis, vertigo, and headache; the latter was categorised into episodic, chronic, and heterocrania (limited to the middle of the head, and thus migraine).[8]

The decline of the Roman Empire and the emergence of a dichotomy between soul and body, also associated with the diffusion of Christianism, caused a reduced interest in the observation of the human body. Throughout the early Middle Ages, what concerns the sick, illness, healing, and therapy appears not structured primarily within the context of rational and scientific coordinates, but rather within the framework of a Christian religious anthropology. In this context, disease (*infirmitas*) was not perceived as an occasional, deplorable moment of imbalance with respect to the norm of health, rather it was seen as the natural condition for humankind after the Biblical fall.[9] Thus, the conception of disease was traced back to a form of supreme divine punishment. The foundation of universities provided the institutional frame for the education of doctors. Physicians were trained on classical texts mediated by the later contribution of Arab-Islamic science.

2 Dante's Knowledge of Anatomy and Physiology of the Nervous System

Dante's *Commedia*, and mainly the *Inferno*, offers several interesting indications of the poet's acquaintance with human anatomy and with certain aspects of the contemporary physiology of the nervous system. The following exam proposes a choice of examples aiming both to illustrate Dante's medical knowledge and to correlate them with modern neurological science.

2.1 *Description of Spasmodic Torticollis or "Manto Syndrome"*

Spasmodic torticollis is a movement disorder and the most common focal dystonia. Its aetiology is unknown, but has been associated with age, trauma, drugs, and genetic conditions.[10] The main clinical feature is a sustained and involuntary muscle contraction that causes an abnormal posture of the neck, and consequently also affects the posture of the head. Around two-thirds of affected patients report pain as a highly disabling symptom. It has been remarked that the disease could have gone underdiagnosed,[11] since a spasmodic torticollis was referred to as *caput obstipum* in ancient literature.[12] The disorder was first described as a pathological condition in the 16th century as "torty colly".[13]

In 1982, a group of Italian neurologists published a case of very high-grade lateral cervical dystonia secondary to brachial plexus damage with improvement within 18 months.[14]

This condition got its name "Manto Syndrome" from the name of the daughter of the Theban soothsayer Tiresias. Manto was a fortune teller punished in Malebolge:

E quella che ricuopre le mammelle,
che tu non vedi, con le trecce sciolte,
e ha di là ogne pilosa pelle,

Manto fu, che cercò per terre molte;
poscia si puose là dove nacqu'io...
(*Inf.* XX, 52–56)

[And she who, yonder, with dishevelled locks / covers the breasts which thou dost not behold, / and has on that side all her hairy skin, / was Manto, who first searched through many lands, / then settled in the place where I was born...].

The rotation of the head determined the gaze directed backwards, so as to recall the torture of the soothsayers in the fourth *bolgia* of the eighth circle of Hell. In *Inf.* XX, 10–18, Dante sees a host of damned crying and advancing in silence, slowly, as in a procession. Looking more carefully, he realises that the figure is distorted and that the face is completely turned back (vv. 11–13: "wondrously twisted each of them appeared between the chin and where the chest begins; for toward his loins his face was turned around" / "mirabilmente apparve esser travolto ciascun tra 'l mento e 'l principio del casso ché da le reni era tornato 'l volto"), so that they are forced to walk backward: "and backward it behooved him to advance, because of foresight they had been deprived" (vv. 14–15: "e in dietro venir li convenia, perché 'l veder dinanzi era lor tolto"). The poet seems to advance a pathophysiological hypothesis of such condition: "By palsy some, perhaps, may thus have been entirely turned around" (vv. 16–17: "Forse per forza già di parlasia si travolse così alcun del tutto"). However, after a brief reflection, he refuses his hypothesis: "but I 've not seen it, nor do I think there ever was one such" (v. 18: "ma io nol vidi, né credo che sia").

Magicians and fortune-tellers are guilty of a grave sin of intellectual presumption. Their insane claim to foresee the future is only allowed to God and to no mortal creature and their punishment is particularly cruel. The retaliation (*contrappasso*) is very clear: since they wanted to push their gaze too far, their heads are turned backwards, and they can only walk backwards.

The poet initially suggests a pathophysiological hypothesis for the damned's condition, attributing it to palsy. However, he quickly abandons his diagnosis, openly admitting that he never observed such a drastic effect resulting from this pathology. Actually, the hypothesis was not correct: in neurological matter, paralysis is a synonym of a function deficit, whereas the aforementioned description reminds a plus function, such as a movement disorder. The extensive encyclopaedic work "On the Properties of Things" (*De proprietatibus rerum*) by the Franciscan friar Bartholomew the Englishman (Bartholomaeus Anglicus), written about 1230–1240, elucidates paralysis in terms that could have been familiar to Dante. Bartholomew defines paralysis as "injury to a part of the human body, with a decrease or deprivation of movement, or sensation, or both" ("laesio partis humani corporis, cum diminutione vel privatione motus, vel sensus, vel utriusque"): it is a *passio nervorum*, as emphasised in the Dante's gloss by Benvenuto da Imola, which *distorquet, dislocat et deordinat collum hominis*. The three types of paralysis defined by Anglicus were strikingly similar to those described by Charles Miller Fisher, a pioneer of stroke neurology in the past century.[15] Miller Fisher identified the so-called lacunar subtypes of strokes, with anatomo-clinical correlations, in particular, he defined five lacunar syndromes, which resembles those described by Bartholomew.

2.2 *Medulla Oblongata and Vital Functions*

Medulla oblongata and the first cervical tract of the spinal cord are fundamental for vital functions. In medulla oblongata there are neurological nuclei that sustain or control respiratory pattern, cardiac rhythm, and consciousness. A lesion located in these parts of the nervous system may cause sudden loss of consciousness (*e.g.* ischemic stroke in basilar artery with ischemia bulb, pons, and mesencephalon) and instant death (*e.g.* lesion in the first cervical metamers of spinal cord with cardio-respiratory arrest). Dante was aware of the function of this part of the central nervous system, and that damage in this area may be frequently lethal, as we can read in various passages.

In Dante's view, the spinal cord is considered the origin of the brain. In the eight circle, ninth *bolgia*, some sowers of discord are punished. Dante observes a damned who advances without his head, which he holds in his hand by the hair, as if it were a lantern. The damned introduces himself to Dante:

E perché tu di me novella porti,
sappi ch'i' son Bertram dal Bornio, quelli
che diedi al re giovane i ma' conforti.

Io feci il padre e 'l figlio in sé ribelli:
Achitofèl non fé più d'Absalone
e di Davìd coi malvagi punzelli.

Perch'io parti' così giunte persone,
partito porto il mio cerebro, lasso!
dal suo principio ch'è in questo troncone.

Così s'osserva in me lo contrapasso.
(*Inf.* XXVIII, 133–142)

[And that thou with thee mayst bear news of me, / know that Bertran de Born I am, the man / who gave the youthful king the ill support. / Of sire and son I mutual rebels made; / Ahithophel by Absalom and David, / with his malicious goadings, did no more. / Because I severed those who thus were joined, / I bear my brain around with me, alas! / severed from its foundation in this trunk; / retaliation thus is seen in me].

Bertran de Born (*c.*1140-*c.*1215) was the lord of the castle of Hautefort, in Perigord, and participated in the disputes of the court of England, supporting Henry Curtmantle (also called the young king), against his father Henry II Plantagenet and his brother Richard the Lionheart. When Richard became king of England, he went over to his side. Since he has divided people so united, he now proceeds with his head separated from the body to represent himself, in his punishment, the faithful application of the law of retaliation (*contrappasso*). The image of the man is particularly powerful, but, above all, indirectly provides some clues of Dante's skills in relation to the anatomy of the nervous system.

The anatomy of the brain had been described by Hippocrates:

The brain is double, in man as in all other animals. A thin membrane divides it in half. Veins stretch out from all parts of the body, numerous and thin, but two thick ones, one from the liver and one from the spleen.[16]

It was also detailed by Aristotle, Erasistratus, and Galen.[17] The latter, in particular, observed that "the texture of the spinal cord is the same as that of the brain from which it starts, but it is more hardened than the brain tissue" and described the medulla "as a river that rises from its source, extends from the brain, sending continuously a nerve channel to each of the parts it encounters, through which both sensation and movement are transmitted".[18]

A passage that highlights Dante's technical knowledge in the medical field, and specifically in neurological functions, appears in the final verses of *Inf.* XXXII, where the poet introduces the character of Count Ugolino. Here, Dante employs the term "nape", *nuca*. This word is a true technicism, derived from the Arabic *nukhā*, originally meaning "spinal cord", which entered the medieval medical language in the environment of the University of Bologna. The term *nucha* was also used by Albert the Great (1206–1280) to indicate not simply the upper part of the neck, but the section of the brain that "administers motion to the entire lower part of the body through the nerves which arise from it".[19]

This notion of *nuca* as the zone of continuity between the two parts of the central nervous system is present in the description of the torture of Count Ugolino, who bites into the skull of the Archbishop Ruggieri:

> e come 'l pan per fame si manduca,
> così 'l sovran li denti a l'altro pose
> là 've 'l cervel s'aggiugne con la nuca
> (*Inf.* XXXII, 127–129)
>
> [And as because of hunger bread is eaten, / even so the upper on the other set / his teeth, where to the nape the brain is joined].

Dante reaffirms the neurological importance of this anatomical zone of the body in the scene of Gianni Schicchi's attack on Capocchio, who, savagely bitten at the "neck knot" (*nodo del collo*), immediately falls to the ground:

> L'una giunse a Capocchio, e in sul nodo
> del collo l'assannò, sì che, tirando,
> grattar li fece il ventre al fondo sodo.
> (*Inf.* XXX, 28–30)
>
> [One reached Capocchio, and so thrust his tusks / into his neck behind, that, dragging him, / he made his belly scrape the solid ground].

2.3 Loss of Consciousness, and the Nape Region

Additional information on the nape region can be found in the seventh *bolgia* of the eighth circle: the thieves punished in this *bolgia* run naked among the snakes and have their hands tied behind their backs by other snakes, often undergoing horrible transformations. Dante meets here Vanni Fucci, his contemporary, a black Guelph, famous for the raids he perpetrated against opposing families. Since he was guilty of stealing the precious objects in the sacristy of the chapel of S. Iacopo in Pistoia, Dante portrays Fucci in a strongly negative way, as a stubborn and remorseless sinner, making him the protagonist of an unusual show. Before declaring his identity and remembering his bestial behaviour, in fact, the sinner is bitten in the back of the head by a snake and, in a few seconds, lights up, burns, and turns to ashes, and then regains his human appearance. He gets up again, stunned, and sighing like someone who falls, without knowing the cause, by the force of a demon that pulls him to the ground or an obstruction of the vital spirits.[20] The point where the serpent bites the damned is of fundamental importance for the purposes of a medical reading of this passage: "there where the neck is to the shoulders joined" (*Inf.* XXIV, 97–99: "là dove 'l collo a le spalle s'annoda"). This is the anatomical region of the nape, at the origin of the spinal cord, which, according to the Galenic theories embraced by Dante, and already mentioned, was in direct continuity with the brain, so much so that the poison inoculated at this point has an immediate effect, promptly causing the symptoms described. The effects of snake bites find precise references in medieval treatises, which provide detailed classifications of the various types of ophidians, with the indication of the names, the effectiveness of the poison, and the consequences of the bite. The change in colour, the astonishment, and the induced sleep have specific connections with the description of the effects of the bite of snakes, classified as *surdi*, in the translation of the *Canon* of Avicenna by Gerard of Cremona: feeling of extreme tiredness, general numbness, invincible sleepiness.

Loss of consciousness is a recurrent event throughout the *Divine Comedy*. Dante seems to have knowledge of signs and symptoms associated with this pathology, as he describes the loss of consciousness in some characters of the *Commedia*, and is also possible that he suffered personally from this disease. From the viewpoint of modern neurology, the causes of LOC (*i.e.* loss of consciousness) can be divided in three groups: (1) structural brain damage; (2) diffuse neuronal dysfunction; (3) psychogenic unresponsiveness.[21] Arousal and awareness are regulated by neurons of the reticular system in the midbrain and in the pons, that connect bilaterally in the thalamus where depart projections to several parts of the cortex, bilaterally.

In the following passage Dante provides two explanations of LOC:

> E qual è quel che cade, e non sa como,
> per forza di demon ch'a terra il tira,
> o d'altra oppilazion che lega l'omo,
>
> quando si leva, che 'ntorno si mira
> tutto smarrito de la grande angoscia
> ch'elli ha sofferta, e guardando sospira
> (*Inf.* XXIV, 112–117)
>
> [And as is he who falls, nor knoweth how, / by demon force, which pulls him to the ground, / or other inhibition binding man, / and who, on getting up again, looks round / wholly bewildered by the great distress / which he has felt, and, as he looks, heaves sighs; / such was that sinner, after he had risen].

Dante therefore provides two explanations relating to LOC: the popular one, which saw a demon-possessed patient in the patient, and the pathophysiological one, which identified the cause of the disease in an "obstruction" (*oppilazione*) of the vital circulation.

To describe the image of the damned who regains human form, Dante uses a very effective expression, characterised by the *stupor mentis* with open eyes, which has been interpreted as a description of both an epileptic fit and a stroke. Commentators have also wondered if Dante wanted to indicate a single morbid event, generated by two different mechanisms, or two different clinical pictures, which have a similar external manifestation: the fall and the amazement. The difficulty of differential diagnosis had already been felt in the past, but Dante tries hard to provide some important clues. First, this "fall" differs from many other cases of sudden falls, which recur in the *Commedia*; sudden unconsciousness, probable tonic-clonic muscular contractions, amnesia constitute the clinical picture of cerebral suffering, the epiphenomenon of which is astonishment with open eyes: "looks round / wholly bewildered by the great distress" (vv. 15–16). Furthermore, as already said, the term "oppilation" is a real technicism, as it was used in medical treatises to indicate the occlusion of a duct, understood as a channel present in the motion nerves or as a blood vessel.

Dante, as mentioned above, therefore provides two explanations relating to epilepsy: the popular one, which spoke of a demon-possessed patient, and the physio-pathological one, which identified the cause of the disease in an obstruction of the vital circulation. In these verses, the scriptural references (Mk 9, 17–27; Mt 17, 14–21; Lk 9, 37–42) coexist with the scientific ones, including the texts of Taddeo Alderotti, who had identified four pathologies, caused by an oppilation that damages the brain: apoplexy and epilepsy, as well as consequences of emotions or external situations. The reference to the demon, in addition to being recalled by the passages of the Gospel, refers to an archaic conception of epilepsy, which was first referred to as organic damage by Hippocrates, but which also remained in vogue in the Middle Ages and in the following centuries, as a popular reading of a pathology that, due to its external manifestations, struck the imagination.

Hippocrates wrote:

> It is thus with regard to the disease called Sacred: it appears to me to be nowise more divine nor more sacred than other diseases, but has a natural cause from the originates like other affections. Men regard its nature and cause as divine from ignorance and wonder, because it is not at all like to other diseases. And this notion of its divinity is kept up by their inability to comprehend it, and the simplicity of the mode by which it is cured, for men are freed from it by purifications and incantations. But if it is reckoned divine because it is wonderful, instead of one there are many diseases which would be sacred; for, as I will show, there are others no less wonderful and prodigious, which nobody imagines to be sacred.[22]

Sacred disease: this is how the ancients, in fact, indicated epilepsy, dismayed by its external manifestations, which could only originate from the superior intervention of a divinity. Hippocrates, on the other hand, addresses the relationship between divinity and nature, between superstition and scientific knowledge, demonstrating the safety of a solid and coherent theoretical system, identifying the cause in the phlegm, which precludes the correct flow of the pneuma in the brain.[23]

2.4 *Fight-or-flight Response*

The response to a powerful threat stimulus driven by the sympathetic autonomic nervous system was first described in contemporary science by Charles Darwin in the 19th century elaborated by Walter Cannon in the early 20th century. Nonetheless, what appears to be the earliest known case is to be found in the Ancient Egyptian work *Story of Sinuhe*.[24] Dante himself explained the response in his encounter with the werewolf:

> Vedi la bestia per cu' io mi volsi;
> aiutami da lei, famoso saggio,
> ch'ella mi fa tremar le vene e i polsi.
> (*Inf.* I, 88–90)

> [Behold the beast, for which I have turned back; / Do thou protect me from her, famous Sage, / For she doth make my veins and pulses tremble].

In it an accelerated heart rate in response to a powerful stimulus is presented, since veins and pulses is a rhetorical poetic form for pulsating veins, which in medieval physiology refer to arteries.[25] The possibility that Dante's fight-or-flight response may indeed be related to an underlying neuropathological condition was considered,[26] whereas a full pathological interpretation of Dante's lucid dream-like descriptions has been provided in recent years (i.e. narcolepsy).[27]

3 Conclusions

In this brief overview of Dante's neurological knowledge, we aimed to offer matter for reflection on how his acute observations concerning human physiology and pathologies can be read in the light of modern scientific knowledge. Far from contextualising them out of their historical frame, the examples discussed in this paper aspire to present insightful views on developing concepts in anatomy and physiology and in Dante's case also provide a valuable snapshot of neuropathologies in the Middle Ages.

Notes

1 Galassi, Habicht, and Rühli, "Dante Alighieri's Narcolepsy"; Lippi, *Dante tra Ipocràte e Galieno*; Lippi, Murru, Galassi, Donell, and Bianucci, "Ophthalmology in Dante"; Bianucci, Lanini, Donell, and Lippi, "Pain and Its Management".

For citations from classical authors, we do not refer to any specific edition. For citations from Dante's *Comedy*, the edition and translation by Langdon has been adopted.

2 Breitenfeld, Jurasic, and Breitenfeld, "Hippocrates".
3 Tsoucalas, Spengos, Panayiotakopoulos, and Papaioannou, "Epilepsy"; Hoppe, "Citing Hippocrates". For the ancient significance of epilepsy at political and societal levels, see also Galassi and Ashrafian, "Has the Diagnosis of a Stroke Been Overlooked". Importantly, some of the descriptions made by Hippocrates of epilepsy in animals have also been reexamined in the light of prion disease, cf. Galassi, Henneberg, and Rühli, "The History of Prion Disease".
4 Gross, "Aristotle on the Brain"; Clarke and Stannard, "Aristotle on the Anatomy".
5 Rocca, "Galen and the Ventricular System".
6 Freemon, "Galen's Ideas on Neurological Function".
7 Guerrero-Peral, de Frutos González, and Pedraza, "Galeata: Chronic Migraine".
8 García-Albea Ristol, "Areteo de Capadocia".
9 Krötzl, Mustakallio, and Kuuliala, *Infirmity in Antiquity and the Middle Ages*.
10 Crowner, "Cervical Dystonia".
11 Jankovic, Tsui, and Bergeron, "Prevalence of Cervical Dystonia".
12 Editorial of *The Lancet* (1978) "Spasmodic Torticollis".
13 Tibbetts, "Spasmodic Torticollis".
14 Disertori, Ducati, and Piazza, "Torcicollo spasmodico"; Disertori, Ducati, Piazza, and Pavani, "Brainstem Auditory Evoked Potentials".
15 Donnan and Norrving, "Lacunes and Lacunar Syndromes"; Miller Fisher, "Lacunes".
16 Hippocrates, *De morbo sacro* 3 (VI 366L), transl. by Adams Francis, "The, Genuine Works of Hippocrates", New York, William Wood & Company, 1886.
17 Clifford Rose, "The Neurology of Ancient Greece".
18 Freemon, "Galen's Ideas on Neurological Function".
19 Gilson, "The Anatomy and Physiology of the Human Body", 22, with reference to Albert's *De animalibus*, I. 2. 11. As Gilson notes, Bruno Nardi was the first to identify the term as a technical word derived from medicine.
20 Riva, Bellani, Tremolizzo, Lorusso, Ferrarese, and Cesana, "The Neurologist in Dante's *Inferno*".
21 Edlow, Rabinstein, Traub, and Wijdicks, "Diagnosis of Reversible Causes of Coma".
22 Hippocrates, *De morbo sacro* 1, transl. by Adams, 334–335.
23 Ureni, "Lo 'stupor mentis' a occhi aperti".
24 Galassi, Böni, Rühli, and Habicht, "Fight-or-flight Response".
25 Galassi, Habicht, and Rühli, "Dante Alighieri's Narcolepsy".
26 Galassi, Habicht, and Rühli, "Dante Alighieri's Narcolepsy".
27 Plazzi, "Dante's Description of Narcolepsy".

Bibliography

Primary Sources

Dante Alighieri. *The Divine Comedy*. Italian text with a translation in English Blank Verse and a commentary by Courtney Langdon. 3 vols. Cambridge: Harvard University Press, 1918, 1920, 1921.

Hippocrates. *De morbo sacro. On the Sacred Disease. The Genuine Works of Hippocrates*, Translated by Francis Adams, vol. 2, 334–346. William Wood & Company: New York, 1886.

Secondary Sources

Bianucci, Raffaella, Iacopo Lanini, Simon Donell, and Donatella Lippi. "Pain and Its Management: Dante's Divine Comedy". *Postgraduate Medical Journal* 99/1170 (2023): 259–264. doi: 10.1136/postgradmedj-2021-140058.

Breitenfeld, Tomislav, Miljenka-Jelena Jurasic, and D. Breitenfeld. "Hippocrates: The Forefather of Neurology". *Neurological Sciences* 35/9 (2014): 1349–1352. doi: 10.1007/s10072-014-1869-3.

Clarke, Edwin and Jerry Stannard. "Aristotle on the Anatomy of the Brain". *Journal of the History of Medicine and Allied Sciences* 18/2 (1963): 130–148. doi: 10.1093/jhmas/XVIII.2.130.

Clifford Rose, Frank C. "The Neurology of Ancient Greece – An Overview". *Journal of the History of the Neurosciences* 3/4 (1994): 237–260. doi: 10.1080/09647049409525617.

Crowner, Beth E. "Cervical Dystonia: Disease Profile and Clinical Management". *Physical Therapy* 87/11 (2007): 1511–1526. doi: 10.2522/ptj.20060272.

Disertori, B., A. Ducati, and M. Piazza. "Torcicollo spasmodico configurante sindrome di Manto, a sospetta etiologia tossica, correlato ad alterazioni dei potenziali evocati acustici del tronco cerebrale. Osservazione per oltre un anno. Terapia con L-5-idrossitriptofano. Riduzione d'alto grado della sintomatologia, normalizzazione dei potenziali". *Annali dell'Ospedale Maria Vittoria di Torino* 25/1–6 (1982): 3–20.

Disertori, B., A. Ducati, M. Piazza, and M. Pavani. "Brainstem Auditory Evoked Potentials in a Case of 'Manto Syndrome', or Spasmodic Torticollis with Thoracic Outlet Syndrome". *Italian Journal of Neurological Sciences* 3 (1982): 359–363. doi: 10.1007/BF02043587.

Donnan, Geoffrey A., and Bo Norrving. "Chapter 27. Lacunes and Lacunar Syndromes". In *Handbook of Clinical Neurology*, vol. 98, edited by Marc Fisher, 559–575. Amsterdam: Elsevier, 2008.

Editorial. "Spasmodic torticollis". *The Lancet* 312/8084 (1978): 301–302.

Edlow, Jonathan A., Alejandro Rabinstein, Stephen J. Traub, and Eelco F. M. Wijdicks. "Diagnosis of Reversible Causes of Coma". *The Lancet* 384/9959 (2014): 2064–2076. doi: 10.1016/S0140-6736(13)62184-4.

Freemon, Frank R. "Galen's Ideas on Neurological Function". *Journal of the History of the Neurosciences* 3/4 (1994): 263–271. doi: 10.1080/09647049409525619.

Galassi, Francesco M., and Hutan Ashrafian. "Has the Diagnosis of a Stroke been Overlooked in the Symptoms of Julius Caesar?". *Neurological Sciences* 36/8 (2015): 1521–1522. doi: 10.1007/s10072-015-2191-4.

Galassi, Francesco M., Michael E. Habicht, and Frank J. Rühli. "Dante Alighieri's Narcolepsy". *The Lancet Neurology* 15/3 (2016a): 245. doi: 10.1016/S1474-4422(16)00029-6.

Galassi, Francesco M., Maciej Henneberg, and Frank J. Rühli. "The History of Prion Disease". *The Lancet Neurology* 15/12 (2016b): 1211. doi: 10.1016/S1474-4422(16)30252-6.

Galassi, Francesco M., Thomas Böni, Frank J. Rühli, and Michael E. Habicht. "Fight-or-flight Response in the Ancient Egyptian Novel 'Sinuhe' (c. 1800 BCE)". *Autonomic Neuroscience* 195 (2016c): 27–28. doi: 10.1016/j.autneu.2016.02.006.

García-Albea Ristol, Esteban. "Areteo de Capadocia (siglo II d. C.) y las primeras descripciones neurológicas". *Revista de Neurologia* 48/6 (2009): 322–327.

Gilson, Simon A. "The Anatomy and Physiology of the Human Body in the *Commedia*". In *Dante and the Human Body: Eight Essays*, edited by John C. Barnes and Jennifer Petrie, 11–42. Dublin: UCD Foundation for Italian Studies-Four Courts Press, 2007.

Gross, Charles G. "Aristotle on the Brain". *The Neuroscientist* 1/4 (1995): 245–250. doi: 10.1177/107385849500100408.

Guerrero-Peral, Angel Luís, Virginia de Frutos González, and María Isabel Pedraza-Hueso. "Galeata: Chronic Migraine Independently Considered in a Medieval Headache Classification". *The Journal of Headache and Pain* 15/1 (2014): 16. doi: 10.1186/1129-2377-15-16.

Hoppe, Christian. "Citing Hippocrates on Depression in Epilepsy". *Epilepsy & Behaviour* 90 (2019): 31–36. doi: 10.1016/j.yebeh.2018.10.041.

Jankovic, Joseph J., J. Tsui, and C Bergeron. "Prevalence of Cervical Dystonia and Spasmodic Torticollis in the United States General Population". *Parkinsonism and Related Disorders* 13(7) (2007): 411–416. doi: 10.1016/j.parkreldis.2007.02.005.

Krötzl, Christian, Katariina Mustakallio, and Jenni Kuuliala. *Infirmity in Antiquity and the Middle Ages. Social and Cultural Approaches to Health. Weakness and Care*. London – New York: Routledge, 2016.

Lippi, Donatella, Chiara Murru, Francesco M. Galassi, Simon T. Donell, and Raffaella Bianucci. "Ophthalmology in Dante: On the Concepts of Sight and Light". *Minerva Ophthalmology* 63 (2021): 1–6. doi: 10.23736/S2785-1265.21.01844-4.

Lippi, Donatella. *Dante tra Ipocràte e Galieno. Il lessico della medicina nella Commedia*. Florence: Pontecorboli, 2021.

Miller Fisher, Charles. "Lacunes. Small, Deep Cerebral Infarcts". *Neurology* 77/24 (2011): 2104. doi: 10.1212/01.wnl.0000410087.34228.7d.

Plazzi, Giuseppe. "Dante's Description of Narcolepsy". *Sleep Medicine* 14/11 (2013): 1221–1223. doi: 10.1016/j.sleep.2013.07.005.

Riva, Michele A., Iacopo Bellani, Lucio Tremolizzo, Lorenzo Lorusso, Carlo Ferrarese, and Giancarlo Cesana. "The Neurologist in Dante's Inferno". *European Neurology* 73 (2015): 278–282. doi: 10.1159/000381541.

Rocca, Julius. "Galen and the Ventricular System". *Journal of the History of the Neurosciences* 6/3 (1997): 227–239. doi: 10.1080/09647049709525710. PMID: 11619860.

Tibbetts, R.W. "Spasmodic Torticollis". *Journal Psychosomatic Research* 15/4 (1971): 461–469. doi: 10.1016/0022-3999(71)90028-6.

Tsoucalas, Gregory, Konstantinos Spengos, George Panayiotakopoulos, Theodoros Papaioannou, and Marianna Karamanou. "Epilepsy, Theories and Treatment Inside *Corpus Hippocraticum*". *Current Pharmaceutical Design* 23/42 (2017): 6369–6372. doi: 10.2174/1381612823666171024153144.

Ureni, Paola, "Lo 'stupor mentis' a occhi aperti di Vanni Fucci (*Inf*. XXIV, 97–118): filigrane mediche e scritturali, in Dante". *Rivista internazionale di studi su Dante Alighieri* 12 (2015): 55–68.

Index

Note: The index does not record the occurrences of Dante Alighieri, mythological and biblical names, and the names of the characters Dante mentions in his writings.

Achilli, E. 185
Adams, F. 199
Akbari, S. C. 121–122, 142
Albert the Great 3–4, 9–11, 15, 18–21, 23–24, 77, 81, 84, 94–96, 102, 105, 107, 111, 121–123, 151, 168, 170, 175, 194
Albumasar 11
Alcmaeon of Croton 190
Alessandro Vellutello 60, 71
Alexander the Great 21
Alfarabi 19, 94, 105, 107, 121
Algazali (Algazel) 19, 94–95
Alhazen (Alhacen) 5, 29, 31, 33–36, 38–42, 56, 66–68, 77–79, 105–106, 110, 112, 114–115, 121–122, 145
Alkindi (Al-Kindi) 19, 33–34, 105, 107, 111
Alpetragius 20
Amico dell'Ottimo 127–128, 130–131
Andrea del Castagno 185
Andrea Lancia 127, 137, 141
Anonimo Fiorentino 5, 127–128, 131–134, 136, 139, 141–143, 145
Antonio Manetti 60
Anzulewicz, H. 24
Appenzeller, O. 185–186
Arba, F. ix, 6
Ardizzone, M. L. 162
Aretaeus of Cappadocia 191
Aristotle 2–3, 8–14, 16, 18–19, 22–23, 30–32, 34, 36–39, 45, 62–63, 67, 79, 95, 105, 107, 110–114, 118, 121–122, 130–131, 136, 140, 144–145, 153, 161, 167–169, 178, 190, 194
Aristotle [pseudo] 38, 42, 131, 144
Arnaldus de Villanova 162
Ashrafian, H. 199
Asín Palacios, M. 184
Atturo, V. 23
Auerbach, E. 57
Augustine (of Hippo) 33, 77, 81, 151, 155, 161–162, 180, 186
Augustine [pseudo] 162
Averroes 3, 10–11, 19, 94, 105, 111, 122, 168–169
Avicenna 4, 11, 19, 33, 84, 94–95, 98, 100–103, 105, 111, 121–122, 151, 162, 168–169, 171, 196
Azzetta, L. 143

Baldi, N. 79
Ballantyne, J. W. 184
Barański, Z. 13, 21, 23
Barbera, G. 185
Barbi, M. 159, 185
Barbi, S. A. 72, 81, 101, 121
Bargellini, P. 70, 81
Bartholomew the Englishman (Bartholomaeus Anglicus) 112, 136, 154, 160–161, 193
Bartolomeo da Bologna 112
Bassermann, A. 69, 74, 81
Beatrice (de' Portinari) 173
Bede (the Venerable) 162
Bellani, I. 185, 199
Bellomo, S. 145
Benvenuto da Imola 5, 18, 21, 127–128, 131–132, 134, 136, 138, 140, 143–145
Bergeron, C. 199
Bernabeo, R. 185

Bernard of Clairvaux 162
Berti, E. 22
Bertran de Born 194
Bianucci, R. ix, 6, 185–186, 198
Boethius 162
Bojita, M. T. 185
Böni, T. 199
Borraccini, M. 186
Bosco, U. 24
Boyde, P. 21, 58, 121, 126
Braid, B. 58
Brambilla Ageno, F. 58, 79, 121
Brandt, T. 42
Breitenfeld, D. 199
Breitenfeld, T. 199
Brigandì, O. 42, 80–81, 123
Brigo, F. ix, 5, 122
Brunetto Latini 154

Caesarius of Arles 180
Cambioli, L. 185–186
Cannon, W. 198
Casagrande, C. 184–185
Casini, T. 72, 81
Castagna, F. 185–186
Castiglioni, A. 185
Cervigni, D. S. 101
Cesana, G. 185–186, 199
Cesari, A. 80
Charlier, P. 185–186
Chessick, R. D. 185
Chiavacci Leonardi, A. M. 72, 130
Chiose cagliaritane 130, 143
Chisena, A. 12
Cianci, N. 185–186
Ciavolella, M. 159, 162–163
Ciccuto, M. 81
Clarke, E. 160, 199
Clement IV (Pope) 106
Clifford Rose, F. C. 199
Clover, W. 160
Cohen, M. R. 42
Colella, M. 81
Constantine the African 159, 169–170
Corsini, G. 184
Corti, M. 72
Cosmacini, G. 159–160
Cotugno, D. 160
Cristoforo Landino 8, 21
Crombie, A. 42
Crone, R. 42
Crowner, B. 199
Cursi, M. 145

Dante da Maiano 159
Darwin, C. 198
de Brosses, C. 68–69, 80
de Frutos González, V. 199
De Renzi, S. 185
De Robertis, D. 22
Delgado, G. 186
dell'Oso, Lorenzo 10
Demaitre, L. 163
Democritus 33
Dewhurst, K. 160
Di Matteo, B. 185
Diana, E. 185
Dino del Garbo 159, 163
Dioscorides 168
Disertori, B. 199
Dod, B. G. 21
Domenico di Michelino 185
Donati, S. 24
Donell, S. T. 198
Donnan, G. A. 199
Drabkin, I. E. 42
Ducati, A. 199
Dumitrascu, D. I. 185
Durling, R. M. 23, 71

Edlow, J. A. 199
Erasistratus 194
Estañol, B. 186
Euclid 30, 32–33, 39–41, 66–68, 130, 133
Euclid (pseudo) 105

Fallani, G. 71
Falzone, P. 23
Fauser, W. 24
Federici Vescovini, G. 120, 122, 142
Fenzi, E. 159, 163
Ferrarese, C. 185, 199
Fiduccio de' Milotti 167–168
Filardo, G. 185
Filosa, E. 23
Finger, S. 160
Fink, J. L. 42
Fioravanti, G. 10, 12, 22–23, 101, 184–185
Fletcher, J. O. 160
Folco Portinari 173
Fontana, S. 185
Foster, K. 121
Forster, S. E. 42
Fosca, N. 20–21, 24, 143
Francesco da Buti 5, 127–128, 132–133, 136, 138–140, 144

Fraticelli, P. 71
Frederick II (Emperor) 167, 184
Freemon, F. R. 199
Frisardi, A. 101
Frixione, E. 160
Fughelli, P. 185

Galassi, F. M. ix, 6, 198–199
Galen 19, 33, 105, 107, 111, 122, 136, 150, 155, 162, 167–170, 178, 190, 194
Galileo Galilei 60
Galli, F. ix, 5, 58, 79, 121–122, 142, 185
Garavelli, B. 72
García-Albea Ristol, E. 199
Gentile da Cingoli 166
Gerard of Berry 159
Gerard of Cremona 11, 168–169, 196
Giacomo da Lentini 122
Giacon, C. 102
Giannattasio, C. 185
Gilson, S. ix, 3, 21–23, 58, 60, 64, 77, 80–81, 121–123, 126, 142, 144, 146, 162, 175, 185, 199
Giorgio Vasari 185
Giotto 168, 185
Giovan Battista Gelli 144
Giovanni Boccaccio 59, 131, 184
Giovanni di Paolo 127
Goethe, J. W. 69, 80
Gomez, E. A. 185
Gomez, W. 24
Gonzo, E. 59
Gorni, G. 101
Graziolo Bambaglioli 71
Gregoric, P. 42
Gregory, T. 102
Grellard, C. 102
Gross, C. G. 199
Guerrero-Peral, A. L. 199
Guglielmo di Maramauro 135
Guido Cavalcanti 13, 85, 90, 129, 151–152, 158–159, 163
Gundissalinus 95

Habicht, M. E. 198–199
Haines, D. E. 185
Haly (Commentator of *Tegni*) 163
Hasse, D. N. 102
Henneberg, M. 199
Henry Curtmantle 194
Henry II Plantagenet 194
Hero of Alexandria 32
Hillard, K. 161
Hippocrates 167–170, 190, 194, 197, 199
Hollander, J. 58, 71, 79
Hollander, R. 20–21, 24, 58, 71, 79, 143
Horace 15
Howard, I. P. 42
Hugh of Saint Victor 156, 162

Iacomo della Lana 5, 127–131, 133, 135–136, 138–140, 143–144
Iannucci, L. 58
Ibn al-Djazzar 159
Ibn al-Haytham *see* Alhazen
Imbach, R. 22
Isaac Israeli 94
Isidore of Seville 162

Jacopo da Pistoia 166
James of Venice 121
James, W. 161
Jankovic, J. J. 199
Jean de Meun 13, 114
Jiménez-Mayo, E. 186
Johannitius 155
John Peckham (Pecham) 29, 34, 56, 78, 82, 106, 112, 114, 121–122, 127
Julius Caesar 183
Jurasic, M.-J. 199

Kemp, S. 102, 160–162
Kepler, Johannes 29
Krötzl, C. 199
Kuuliala, J. 199

Lacarra Lanz, E. 163
Lange, C. 161
Lanini, I. 198
Lansing, R. H. 58, 79, 121
Ledda, G. 122, 142
Lejeune, A. 42
Leonardo Bruni 60
Leonardo da Vinci 146
Leopardi, G. 52
Li Calzi, E. 185
Lička, L. 121, 145
Lindberg, D. C. 120–122, 142, 145, 162
Lippi, D. x, 6, 184–186, 198
Lo Presti, M. 185
Locatelli, L. 186
Loconsole, M. 23

Lombroso, C. 176, 185
Lorenzo Ghiberti 122
Lorusso, L. x, 6, 166, 185, 199
Lowest, J. L. 163
Lucan 23, 39, 80
Lucchini, R. 185
Lucretius 32, 37–40, 42, 66, 68, 79, 81
Lunedei, A. 185
Lustig, R. 185

Madame de Neuilly 68, 80
Madame de Staël (Anne-Louise Germaine Necker) 81
Malato, E. 143
Malgieri, M. E. 23
Mandel'štam, O. E. 1, 7
Mandelbaum, A. 71, 101
Manfred of Sicily 167, 184–185
Manzini, F. 186
Maraldi, E. 185
Marcacci, M. 185
Marcus Aurelius (Emperor) 190
Marinini, M. 186
Martinez, R. 71
Martini Bonadeo, C. 121
Martins, P. N. 185
Masaccio 168
Maurina, M. 186
Maximilien Misson 81
Mazzucchi, A. 143
McGinnis, J. 121
Michel de Montaigne 81
Miller Fisher, C. 193, 199
Minio-Paluello, L. 21–22
Momigliano, A. 72
Mondino de' Liuzzi 167–168, 185
Montefusco, A. 23
Moore, E. 21
Mosca, M. 185
Moses Maimonides 95
Murru, C. 198
Musa, M. L. 121, 159
Mustakallio, K. 199

Nardi, B. 18–20, 95, 123, 159, 163, 199
Narducci, E. 122
Nespoli, A. 186
Niccoli, G. 58
Norrving, B. 199

Oliva, G. 59
Olry, R. 185
Ottimo Commento 5, 71, 127–128, 130, 133, 136, 138, 140, 143, 146

Padoan, G. 143
Palazzo, A. 24
Panayiotakopoulos, G. 199
Panti, C. x, 4–5, 58–60, 79, 121–123, 142
Paolo Uccello 168
Papaioannou, T. 199
Paracelsus 168
Paravicini Bagliani, A. 121
Parodi, E. 72, 81
Parronchi, A. 77, 81, 114, 121–122
Pasi, R. 185
Pasquini, E. 71
Paul (Saint) 91–92
Pavani, M. 199
Pedraza-Hueso, M. I. 199
Pegoretti, A. 22, 123
Perciaccante, A. 185–186
Peter Lombard 12
Peter of Spain (Petrus Hispanus) 151, 163, 170
Petrocchi, G. 58, 79, 101, 144, 160
Piazza, M. 199
Piccolino, M. x, 1, 4, 42, 60, 79–80, 142, 146
Pier delle Vigne 184
Pierre de Limoges 112
Pietro Alighieri 130, 136, 143
Pietro d'Abano 95
Pietro da Pisa 71
Pisoni, G. 145
Plato 12, 30–34, 107, 169
Plazzi, G. 177, 185, 199
Porro, A. 185
Premuda, L. 185
Propertius 15
Ptolemy (Claudius) 11, 29, 31–36, 38–42, 66–67, 81, 105–106
Pucci, J. 23

Quaglino, M. 146
Quaglio, A. 71

Rabinstein, A. 199
Raynaud, D. 121–122, 145
Reggio, G. 24
Richard of Middleton 12
Richard of Saint Victor 16
Richard the Lionheart (king) 194
Ricklin, T. 24
Riesman, D. 185
Riva, M. A. x, 6, 185–186, 199
Robert Grosseteste 34, 56, 76, 106, 121
Robiglio, A. 22

Rocca, J. 199
Roger Bacon 29, 34, 56, 78, 106, 112, 114, 121–122, 145, 151
Rogerius of Salerno 170
Rolando Capelluti (Capezzuti) 170
Ronchi, V. 42
Ross, W. D. 42
Rossetti, D. G. 70
Rossetti, G. 70, 81
Rossi, L. C. 160
Ruffino, L. 185
Ruggieri (Archbishop of Pisa) 195
Rühli, F. J. 198–199
Ruskin, J. 81
Rutkin, H. D. 24

Sabra, A. I. 42
Sandolo, A. 185
Sandor, V. 185
Santagata, M. 60
Santi, F. 58
Sapegno, N. 53, 59
Sbacchi, D. 23
Schoonheim, P. L. 22
Scordo, F. 185
Seneca 36
Sentíes-Madrid, H. 186
Sermonti, V. 71
Singleton, C. 71, 130, 160
Siraisi, N. 185
Sironi, V. A. 185
Smith, A. M. 42, 105, 115, 120–121, 142, 145–146
Smith, C. U. 160
Smith, J. L. 186
Socrates 95
Sorabji, R. 121
Spengos, K. 199
Stabile, G. 23
Stannard, J. 199

Tabarroni, A. 166, 170, 184
Tachau, K. 120
Taddeo Alderotti 159, 167, 197
Tarabella, V. 185
Tarud Bettini, S. 122
Tavoni, M. x, 4, 58–59, 80, 101
Theophrastus 30, 38
Thomas Aquinas 3, 10, 18, 20, 59, 77, 95, 102, 105, 111, 121–122, 151, 161, 163, 168, 182
Tibbetts, R. W. 199
Tobino, M. 166
Tomazzoli, G. 123
Tomba, P. 185
Tommaseo, N. 71
Tonelli, N. 159–163
Toynbee, P. 22
Traub, S. J. 199
Travaglia, P. 121
Tremolizzo, L. 185, 199
Tsoucalas, G. 199
Tsui, J. 199

Ureni, P. 186, 199

Varchi, B. 146
Varotto, E. xi, 6
Varrenti, M. 185
Vasoli, C. 20, 22–23
Vasta, E. 101
Verbeke, G. 102
Verstraten, F. A. J. 37, 42
Vesalius (Andreas van Wesel) 168
Viganò, A. 185
Vincent, J.-D. 175–176, 185
Virgil 14–15, 23, 39, 50
Volpi, M. 143, 145
Vuillemin-Diem, G. 22

Wade, N. J. xi, 1, 3, 42, 60, 80, 120
Webster, C. 42
Widmer, D. A. J. 177, 185
Wijdicks, E. F. M. 199
William of Moerbeke 11
William of Saliceto 170
William Shakespeare 59
Witelo 29, 34, 56, 78, 106, 112, 114
Witkovsky, P. 58
Wlassics, T. 45, 54, 56, 59, 79
Wolfson, H. A. 160

www.ingramcontent.com/pod-product-compliance
Lightning Source LLC
LaVergne TN
LVHW021127300926
849160LV00010B/890